M

JUL 2 4 2008

UPGRADE ME

Also by Brian Clegg

The Global Warming Survival Kit
The God Effect: Quantum Entanglement, Science's Strangest Phenomenon
A Brief History of Infinity
The First Scientist: A Life of Roger Bacon
Light Years: An Exploration of Mankind's Enduring Fascination with Light

UPGRADE ME

Our Amazing Journey to Human 2.0

Brian Clegg

St. Martin's Press ❦ New York

www.stmartins.com

Library of Congress Cataloging-in-Publication Data

Clegg, Brian.
    Upgrade me : our amazing journey to human 2.0 / Brian Clegg.—1st. ed.
        p. cm.
Includes bibliographical references.
ISBN-13: 978-0-312-37157-9
ISBN-10: 0-312-37157-8
1. Human evolution—Popular works. I. Title.

GN281.C544 208
599.93'8—dc22   2008018000

First Edition: August 2008

10   9   8   7   6   5   4   3   2   1

For Gillian, Chelsea, and Rebecca

Contents

Acknowledgments

My thanks to my agent, Peter Cox, and my editor, Michael Homler, for making this book possible.

Introduction

Sometime last year I was walking through a field with my dog, a scatterbrained but affectionate golden retriever. Our walks together are recreational, but I do regard them as part of my work routine. Not only do they provide exercise—an essential with the sedentary lifestyle of a writer—but they make a positive contribution to my creativity. If I need an idea, there is nothing better than taking a walk through the countryside for stimulating new thoughts. In this case I wasn't looking for anything particular, but I was to be given the inspiration for this book.

It was a cold day and I was underdressed for the weather. In a short-sleeved shirt, I was shivering. My sneakers had picked up too much moisture from the wet grass and I squelched as I walked along. When passing through the fence from one field to the next, I managed to brush against a rampant clump of nettles, stinging both my arms. But the dog, with her thick fur coat and hard padded feet, was impervious to both the weather and the aggressive vegetation. In evolutionary terms, she seemed much better prepared to survive what nature could throw at us than I was.

My first thought was to wonder why—why human beings are so magnificently badly equipped to cope with the discomforts and dangers of the natural environment. We know that our ancestors and those of the great apes all had good, thick coats of protective fur, just as the apes still do today. (I'm sure it shouldn't be necessary to point out that present-day apes such as chimpanzees and gorillas aren't in any sense our ancestors, but it's a mistake that's still often seen.) It seemed counterintuitive that the early humans should have lost that helpful fur.

Of course, it's a misunderstanding to think that evolution has our best interests in mind. Evolution doesn't have a mind or any concept of what is good and bad for us. Evolution usually works by gradual selection of subtle variants that enhance the survival and reproduction capabilities of individual members of species. It doesn't take an overview and think, "That's good; I'll keep that." Even so, it seemed unlikely that there was any evolutionary benefit in losing the warmth and particularly the protection of that natural fur coat.

Some do still believe that there must have been a benefit from becoming relatively hairless. Fashion writer Julian Robinson comments that many historical philosophers noted our natural naked state: "The fact that they, like us, had only a thin covering of skin, which, although fragile, served to keep their blood in and water out, indicated in evolutionary terms that this must have given these people, and our even earlier progenitors, some special advantage as a species that enabled them to survive or thrive." Robinson goes on to ask why so many human societies have abandoned this "obvious evolutionary advantage" of having exposed skin by wearing clothes.

Unfortunately, this stance implies a fundamental misunderstanding of evolution. It suggests that evolution is not the blind watchmaker of Richard Dawkins's eloquently titled book but a directing force. In his book, Dawkins was referring to the argument for the existence of a creator used by the philosopher William Paley. This eighteenth-century cleric made the suggestion that should we be walking along and observe a watch, with all its complex workings, lying on the ground, then even if we had no foreknowledge, we could deduce the existence of a creator. The watch is clearly designed with a purpose—a function—in mind, and that implied that there was a designer.

Evolution based on natural selection does not require a designer. The astrophysicist Fred Hoyle, best known for explaining how the elements were formed in the stars, once said that the chances of one of the complex life-forms we have on Earth evolving as a result of pure chance were about the same as those of a strong wind blowing through a scrap yard full of junk and assembling a working 747. But for all his scientific expertise, Hoyle did not understand evolution. It isn't a designer, but neither is it random chance. Evolution involves natural selection—so the complex evolves from the simple, step-by-step, when each individual step provides benefit—but evolution has no grand plan, no direction in "mind." Hence Dawkins's *blind* watchmaker.

Just because evolution deals us a set of cards doesn't mean that everything we receive in our genetic hand is directly beneficial. There doesn't have to be an "obvious evolutionary advantage" just because we have a certain trait. A particular characteristic is just as likely to be a side effect of another evolutionary development as it is to be something that gives us benefit directly. For example, many bird wings are easily snapped, because the bones are hollow. Having weak bones isn't a good thing in itself; it's just necessary to enable the bird to fly.

I was eventually to find an answer to why I was suffering as I walked through the field, an explanation of why it made evolutionary sense to lose our fur as an unwanted side effect of another change, in the work of zoologist Clive Bromhall, which I will return to in the first chapter. But soon the question of why we aren't better protected naturally, a question that started well to the front of my mind thanks to those nettle stings, was put aside.

I was struck by the realization that despite being so feeble, we are awesomely good at coping with anything the environment can throw at us. I might not be well suited to walking unaided through a cold field full of nettles, but with the right clothes it's a trivial task. With special protection, I can go plenty of places even the hairiest creature couldn't survive. Human beings aren't limited to the physical (or mental) capabilities than we are born with. Thanks to our remarkable brains, we instead upgrade ourselves to overcome our limitations.

When our distant ancestors became human they gained the ability to think into the future. To see how things could be different. To *desire* change. And very soon that longing for change was focused on themselves. The bodies they were born with were just starting points; they could do so much more. It was not long before they did. . . .

1.

Beyond Biology

Man can improve himself but never will himself be perfect.

—W. H. Auden, "In Time of War"

Around one hundred thousand years ago our distant ancestors went through the final changes that made them into modern humans. From the traditional biological viewpoint, that was the end of our evolution to date. We are the same biological species now as they were back then. There have been plenty of tiny changes at the genetic level, but as a species we are essentially the same. We have the same potential for physical strength, for longevity, for attracting the opposite sex, for thinking, and more. Yet no one would confuse one of our ancestors, lifted out of time, with a modern person off the street. In the intervening hundred thousand years—no time at all on an evolutionary timescale—we have moved beyond our biology, thanks to our phenomenal ability to upgrade ourselves.

Those many thousands of years ago, our predecessors had undergone huge evolutionary changes from the common ancestor they shared with chimpanzees and the other great apes. The prehumans had lost most of their hair, leaving a delicate, thin skin exposed. They had shifted from using a four-legged gait to walking upright. Their brains had grown out of all proportion with their bodies, leaving them bulgy headed and top-heavy (no doubt very unattractive features to their predecessors). Their mouths had become smaller, making their teeth less effective as a biting weapon. The big toe had ceased to be an opposing digit that could be used to grip a tree branch. As a result of this, their feet had lost much of their practical functionality.

Taken together, these alterations seem to be the entire opposite of everything we expect from natural selection. They made the prehumans significantly more vulnerable to attack by predators. Their naked, unprotected skin was pathetically easy for claws and teeth to rip through. Compared with the smooth, four-footed pace of the other apes, their tottering movements on two legs were slow and clumsy—even a rabbit could outrun this strange unstable creature. The adaptations that came through in prehumans don't seem to make any sense. Or at least, they don't make any sense until they're seen as side effects. Seen alone, as evolutionary changes in their own right, they reduced the chances of survival, but put them alongside the change of behavior that triggered them and they were an acceptable price to pay.

These physical modifications of prehumans were an indirect result of an environmental upheaval. As the global climate underwent violent change, our ancestors were pushed out of the protective forests into the exposed world of the savannah. Facing up to coldly efficient predators, they were forced to change behavior or to become extinct. Back then, most prehumans could not function well in large groups. This is still the case with most of our close relatives. The chimpanzee, for example, is incapable of forming large, cooperative bands. Get more than a handful of males together and the outcome is bloody carnage as battles for supremacy break out.

The prehumans who first straggled onto the savannah around 5 million years ago were probably much the same. But the fast, killing-machine predators of the day—from the terrifying saber-toothed dinofelis and the lion-sized machairodus to the more familiar hyenas—made sure that things would change. The most likely prehumans to survive were those with a natural tendency to cooperate. Our ancestors began to operate in larger and larger groupings, giving them the ability to take on a predator and win, where a small roaming band would be torn to pieces. And this change of behavior brought with it as a side effect all the physical oddities that we observe in modern man.

The characteristics that repressed in-group aggression and enhanced the ability to cooperate are typical of juvenile apes. Our primate cousins' inability to function in large groups only appears with maturity. The individuals among our predecessors who were more likely to survive on the savannah, those with the immature ability to get on with their fellows rather than tear them to pieces, were also the least physically developed. The eventual outcome was lack of hair on most of the body, a large head, a small mouth—even the upright stance—all features of the early part of the primate life cycle that have normally disappeared by the time they are mature.

This mechanism of selecting for cooperative behavior and getting an infantlike version of the animal as a side effect is something humanity has since managed to produce repeatedly in its domestic animals. The dog, for example, has much more in common with a wolf cub than with the mature wolf that it was originally bred from. This is not just a matter of theory. In a fascinating long-term experiment between the 1950s and the 1990s, Russian geneticist Dimitri Belyaev selectively bred Russian silver foxes for docile behavior and showed just how early man managed to turn the wolf into a dog.

Over forty years—an immensely long experiment, but no time at all in evolutionary terms—the fox descendants began to resemble domesticated dogs. Their faces changed shape, becoming more rounded. Their ears no longer stood upright but drooped down. Their tails became more floppy. Their coats ceased to be uniform in appearance, developing color variations and patterns. They spent more time in play and constantly looked for leadership from an adult. As they became more cooperative, they took on the physical appearance and the behavior patterns of overgrown fox cubs.

In the process of becoming more cooperative, more infantile ("neotenous" in the scientific jargon), the prehumans had a physical resemblance to a modern human being for many hundreds of thousands of years, but still something was missing. They remained purely animal in their reaction to their surroundings. But with the second breakthrough around one hundred millennia ago, something new, something unique in terrestrial biology, came about. The physical changes that had produced an infantile grown-up ape made possible one further change, the most dramatic of all.

Zoologist Clive Bromhall has described this as a partitioning of the brain, enabling the early humans to simultaneously experience an internal and an external world. Our ancestors began to scrawl pictures on rock walls, to represent in images animals that weren't present. They drew events that took place in the past or might happen in the future. Something had changed in the way their brains functioned, something that opened up the ability to see beyond the now. At the same time as reacting to the world about them, these transformed creatures were able to deal with "what if?," to dream, to plan, to anticipate. They had become conscious.

Watch a TV documentary set in an African game park and the response of prey animals like a herd of gazelle to the presence of predators seems unbelievably strange from the human viewpoint. If a lioness is lying at the edge of the herd, watching intently, picking out a target, this fearsome predator is likely only to be eyed briefly, if nervously, by its potential victims before the gazelle return to cropping the coarse grass. We would be thinking, "I've got a problem here. That lioness could hurt me or even kill me. I think I'll sneak away, just in case. Or at least I'll make sure there's a fatter, slower gazelle between me and the lioness." But this ability to project into the future, to be aware of potential circumstances and analyze consequences, isn't present in the gazelle. It is only when the attack commences that a flight response is triggered.

There are clear survival benefits from being able to consider what might be as well as what is. It gave humans the ability to assess risk, to make decisions based on what might happen, rather than reacting solely to the immediate threat. (Interestingly, we're still not very good at assessing risk. The requirement to calculate probability, something very new to us in biological terms, remains very challenging. If you doubt this, take a walk around a casino.) Seeing beyond the now brought us literature and religion, science and civilization. Yet perhaps the greatest benefit that would come from this change was the realization that we ourselves could become different in the future. Thanks to the ability to ponder what might be, our predecessors were able to think, "I want to be different from the way I am now," kick-starting the urge to upgrade the human form.

The result was something biologically unique. Human beings began to turn themselves into something new, not through the painfully slow process of natural selection but by our own intervention. Genetically, nothing happened. Let's say it again: we belong to the same biological species as our predecessors one hundred millennia back. But this misses the point. Evolutionary change occurs because of *natural* selection—new features that have a survival benefit tend to stay with the species. We might not have gained features naturally, but we have done so *unnaturally*—our desire to improve has driven us to upgrade continuously.

In the 1980s, science fiction writer and mathematician Vernor Vinge devised the concept of the Singularity. Vinge predicted that "within 30 years, we will have the technological means to create superhuman intelligence. Shortly thereafter, the human era will be ended." This idea was picked up by futurologist Ray Kurzweil: "[W]e can reliably predict that, in the not too distant future, we will reach what is known as The Singularity. This is a time when the pace of technological change will be so rapid and its impact so deep that human life will be irreversibly transformed." But Singularity enthusiasts have been misled by biological parallels, expecting the change to be a single, clear leap. In fact, the change is already upon us and has been for thousands of years. We hardly had time to realize that we were human before beginning to make ourselves something more.

Biological species are clearly differentiated. The earlier members of the genus *Homo,* such as *Homo erectus,* were distinct species, clearly distinguished in a biological sense from *Homo sapiens.* This is quite different from our own upgrades, which consist of many, many small additions and improvements. We have been able to do this because of a fundamental difference from the blind watchmaker of biological evolution. Our self-improvement is a truly directed enhancement. Once you have a driving force behind selection and innovation, timescales are transformed.

Just as computer software tends to gain lots of small changes, it would be more accurate to describe us as human version 4.1231.22 (or some other arbitrary version number) than the neat but misleading Human 2.0 that the concept of the Singularity suggests, though for practical purposes I will continue to refer to Human 2.0 as a description of our self-enhanced state.

My academic background is in Operations Research. This discipline, originally devised by the likes of Patrick Blackett in the United Kingdom and William Shockley in the United States to deal with military operational problems, takes a hard scientific approach to big questions in soft subjects. It was this Operations Research background that made me realize that by limiting our view to traditional biological measures of evolution we were missing what was truly happening. What was important was not the means of change but the drivers and the outcome: the added functionality that we had given ourselves.

This process of upgrading began in early humans through the natural pressure of survival rather than as a result of any great vision of building a superhuman future. Our early ancestors weren't looking to create Human 2.0; they wanted to live longer, to become more attractive to the opposite sex, to be better able to defend themselves, to make the most of their brains, and to repair damaged bodies. Opposing these urges was the whole natural universe, indifferent to our desires but often hostile. It wouldn't be easy.

The natural evolutionary answer to a hostile environment would be blind trial and error, resulting in tiny incremental changes over millions of years. Our big brains allowed us to short-circuit this route, and the outcome is ever more startling. Whereas making the body more attractive now means personal styling, it could soon involve glowing firefly hair and tattoos that show living, moving images. Some gerontologists believe that there are people living today who will still be alive in 2150. Implanted devices might not only replace damaged organs but also be linked directly to the brain to enhance memory or to allow direct person-to-person communication.

The high-tech possibilities of the future are stunning, but we shouldn't let them blind us to the importance of the simpler upgrades that are a fundamental part of our lives every day. These were the changes that rapidly shifted early humans away from their natural evolutionary position and that continue to have a huge impact on our comfort and survival. Recently I was walking a national trail on a blazing hot day. For an unmodified *Homo sapiens* this (recreational) journey would be sheer madness. It would just not have been possible. There was no water along the trail—the ground was bone dry. But I was able to keep going hour after hour because I carried water with me. My water bottle achieved what nature took millions of years to develop in a camel—the ability to carry enough water to survive an arid terrain. An unremarkable fifty-cent plastic bottle replaced a huge biological transformation.

We have gained massive advantages from our ability to run our unnatural evolution at ultrahigh speeds. But there are risks as well. The concern is not the creation of a science fiction monster. Natural selection pressures leave such unsustainable developments as dead ends. But our development speed, as we have seen of late with global warming, can put the whole environment at risk. The danger from upgrading is not so much turning ourselves into freaks as the fallout that our self-improvement generates.

Even so, most would agree that the benefits far outweigh the risks. It's easy to be blinded by high-tech predictions into missing what we've already got: the most amazing thing is not what may happen but what is already here. There's no need to look forward with Vinge and Kurzweil to Human 2.0. It's you.

To better understand this unnatural evolution and its implications, we have to first see what the drivers were that kicked us into action. Several factors would power this enthusiasm for change—and one of the earliest of these was the awareness of death.

2.
Cheating Death

I don't want to achieve immortality through my work. I want to achieve it through not dying.

—Woody Allen

The realization "I am going to die" must have come as a huge shock to early humans. Animals can be aware of the deaths of others, but there is no evidence that they realize that they themselves will perish. The ability of *Homo sapiens* to consider its future made it possible to see the inevitability of personal death. This awareness proved an early trigger for human upgrades.

Perhaps the first, and simplest, of responses to realizing the reality of our own mortality mirrors my thoughts while out walking. In developing into human beings we had lost our natural protection of fur to hold back the cold and to defend us from the impact of cuts and scratches produced by the hostile environment we passed through. The development of clothing has enabled us to go far beyond the protective capabilities of a simple coating of fur and allowed us to operate in a wide range of environments. However, perhaps surprisingly, all the indications are that clothing came about as much, if not more, to indicate social status and for sexual allure than to provide warmth and protection, so I will come back to garments in the next chapter.

One aspect of clothing, however, deserves its place here without question. There are some clothes that really don't have any significance other than to defend us from damage by the natural environment or, even more important, from the most dangerous thing on the planet—mankind. Armor may sometimes be decorative or ceremonial, but there can be little doubt about its main function.

For thousands of years the best answer to the need to protect the body from attack, all the way from prehistory to the end of the fifteenth century, was either quilted fabric or leather. Even a millennium after the collapse of the Roman Empire (where metal plating and chain mail was used to some degree), these apparently weak forms of armor were the standard for ordinary soldiers. Leather (and particularly quilting, something we would now associate more with bedspreads than warfare) may not seem ideal protection from attack, but it was light and wearable, easy to work, and capable of turning away relatively blunt sword blows. These two materials would be the mainstay of protective wear until the next technological revolution brought in the introduction of iron armor.

Although the Iron Age itself dates back up to two thousand years earlier, it was only around the time of the Norman conquest of England in 1066 that iron armor, in the form of chain mail and arrays of small metal plates, really came into vogue. There was evidence of the use of iron armor beforehand. Charlemagne, king of the Franks from 768 and the conqueror of an empire that included much of Europe, is described as having an iron breastplate, helm, and shield, with iron plates providing protection elsewhere. The life of Charlemagne written around 883 by the Monk of Saint Gall also relates that Charlemagne had iron greaves but considers this not so exceptional, "for the greaves of all the army were of iron." Greaves, named after the medieval French word for the shin, were protectors for the legs, a type of armor dating back to the ancient Greeks.

Early on, plate armor was often in the form of metal scales, rather like those of a fish, sewn onto a more traditional leather or quilted protector. Chain mail—flexible metal clothing made of interlinked rings of iron—was to become very common but initially would have been exotic. Its flexibility made it hugely appealing compared with metal sheets, but when first produced it was very expensive. The technology to make wire by drawing out heated metal wasn't developed until the 1300s—until then, each ring had to be individually made by a blacksmith heating a bar of iron, then beating out a portion of it into a long strip that would be coiled into ring form and riveted to its neighbors, very time-consuming and costly.

A typical outfit for a soldier at the time of the Norman Conquest (1066) would be a tunic, effectively a long linen shirt worn to just above the knee, then a gambeson (other exotic names for this garment being a wambais or an aketon), which was a second, quilted tunic, again around knee length. The basic foot soldier would make do with this for protection, but a wealthy soldier—typically a knight—would go on to add extra layers. Next would come the hauberk, a chain mail shirt of similar length to the tunics. Where for the foot soldier the versatile gambeson was the sole armor, for the knight it acted to avoid too much damage from the rings of the chain mail being forced into the flesh by a blow. The hauberk would have slits on the front and back so that the wearer could sit on horseback. (Riding a horse was essential because of the weight of the full chain mail. It was too much of a burden for a foot soldier.)

Over the gambeson, some knights also wore a *plastron de fer*—armor made up of metal chest and back plates—and finally a surcoat, also called a jupon, a cloth garment that provided an outer cover to reduce the chance of the chain mail rusting and that could also be used to display any colors that were required. Some form of recognizable decoration, eventually becoming the "coat of arms," was necessary, as the knight's headgear made it difficult to recognize who he was. Legs would be protected in much the same way as the body—cloth or leather bound around them for the common soldiers and for the knights (whose legs were particularly susceptible to attack when they were raised up on horseback) either mail stockings or leather garments studded with metal.

Head protection around the time of the Norman Conquest, as with modern battle helmets, gave the advantage of visibility with the disadvantage of leaving the face largely unprotected. The helmets of the time were relatively tall, cone-shaped headgear made of four triangular pieces of metal riveted together, probably fixed on with leather straps. A degree of protection was often given for the face by providing a strip of metal that came down over the nose, leaving the eyes and cheeks open.

Over the next two hundred years, this helmet evolved into the helm, the headgear we tend to associate with a medieval knight, giving whole-head protection. While the later helms used in jousting were supported on the shoulders, these early heavy metal helmets were supported on the head with a padded cap beneath to protect the wearer, putting extreme strain on the neck muscles.

Over the next three hundred years, plate armor gradually replaced chain mail. First on the legs—so exposed on the horse-borne knight—then as an expanded version of the *plastron de fer,* giving full chest and back protection from sword blows, lances, and arrows, and finally adding all the complex jointed sections to deal with arms, shoulders, elbows, and so on, to give the full plate suit of armor as we imagine it today. Although plate armor was extremely heavy, it was easier on the wearer than full chain mail, as all the weight of the chain mail hung from the shoulders, while the plate was strapped onto various different parts of the body, so the weight was spread around and was easier to bear.

This was not the end of the development of armor. Whatever the military age, there is a literal arms race between the makers of defensive mechanisms and those building offensive weapons. All through the ages these two different enhancements of our natural evolved state have been in conflict—armor to improve on our thin skin that offers little protection to the fragile flesh and weapons to improve on our relatively unimpressive teeth, fingernails, and fists. Plate armor was devised to protect against the sword, lance, longbow, and crossbow—but then came gunpowder. What happened next is often oversimplified by saying that guns rendered armor obsolete. In reality, the weapon builders didn't have it so easy.

Crude guns came into use toward the end of the fourteenth century. The response was to use thicker armor plate and even, in the seventeenth century, to switch from iron to steel. On the whole, armor kept up with the weaponry until the musket with rifling came along, by which point no human-borne metal armor could stand up to the penetrating power of a bullet. To provide adequate protection, armor would have to have been so thick that it wasn't possible to wear it, and full armor would only come back into fashion when it could instead be driven around in, on the outside of an armored vehicle.

Before its demise (by which time it was more decorative than functional) plate armor evolved into a hugely sophisticated piece of engineering. It had to be light enough for the knight to fight in it, well jointed enough to move in, yet thick enough to withstand the strongest blows. Although much medieval armor looks as if it were designed for its appearance, in fact many of the elegant features are there for tried and tested practical reasons. If a man was struck by a lance, for example, which has huge potential for damage from the leverage given by its long pole, the armor had to guide the point away from the place it initially impacted, using curvature that avoided giving the point anywhere to stick but instead forced it to slip harmlessly off the suit.

It was realized surprisingly late that the horse, such an essential for the armored knight to be able to fight, was more of a target than the rider. Bigger than a human being, so easy for bowmen or spear throwers to hit, the horse was left entirely unprotected by the Normans, who seemed to have forgotten the example of the Romans, who had given their horses a form of mail to protect them. It was only in the thirteenth century that horses were given padded garments and then armor. This would be the closest predecessor to a modern armored vehicle, though of course the armor of a car is as much intended to protect the human cargo as it is to make sure that the vehicle itself isn't damaged.

In the present day, armor has returned to its roots as technology has enabled us to produce materials such as Kevlar (a synthetic fiber originally developed to replace steel belts in car tires) that are as light as leather or quilting but that can provide effective protection against the projectile weapons that made plate armor obsolete. Unlike some weaponry, which has become more detached and impersonal, armor remains very much a part of the person, a protective exoskeleton to help overcome the deadly weakness of our thin skin.

Although the defensive armor of the military may be the first that springs to mind, we shouldn't forget what is effectively the oldest form of armor, one intended to ward off the extremes of nature. We now have the ultimate examples of protective clothing that can keep practically any hostile environment at bay. In previous centuries the most dramatic use of clothing for survival was that of the diving suits worn for underwater work, but with the twentieth century these lost their preeminence to the suits used for space walks. Life was never intended to cope with the extremes that space can throw at it. There's no air to speak of. The temperature is impossibly low. It's quite literally like nothing on Earth. Yet astronauts regularly take jaunts in space protected only by this very specialized clothing.

For most of us, the closest we come to experiencing the need to be protected from the vacuum of space is in science fiction movies, but Hollywood has been responsible for some magnificently bad portrayals of what happens when space suits fail. The most ludicrous example is in the otherwise quite intelligent 1990 Arnold Schwarzenegger movie *Total Recall*, based on a Philip K. Dick story, where human beings, expelled from the protected environment of a city on Mars, inflate grossly before their heads explode messily over the surface.

Mars does have a slight atmosphere (around 1 percent of Earth atmospheric pressure), but even in space this sort of inflation and explosion caused by low pressure isn't going to happen. There would be some discomfort as gas escaped from body cavities, but there is no danger that your head would inflate like a balloon.

It's just possible the makers of the movie had in mind another "dying in space" myth that has a little more science behind it than their special effects. Liquids boil at lower temperatures as air pressure drops, so some people have assumed that your blood would boil in your veins if you were ejected into space—not a pleasant way to go. Certainly blood *would* boil in the practically nonexistent pressure of space at body heat if it weren't confined, but according to NASA the containing pressure of your skin and circulatory system is enough to avoid this horrible end.

Another common misapprehension is that humans would instantly freeze to death in the "absolute zero" of space. This idea misses the obvious benefits of a vacuum for keeping heat in. The reason a vacuum flask keeps coffee warm (or chilled things cold) is that it's a great insulator. Heat doesn't pass easily through empty space. The extreme low temperature of space, technically around −450°F, is only a measure of the low energy of empty space. Your body will radiate heat as infrared, but it's not a great radiator, so you will only gradually grow cold. Again, there will be discomfort because exposed fluids, like those that lubricate the eye, will boil away. But death in space is likely to come from good old-fashioned asphyxiation—lack of air—which will take a number of seconds.

NASA even has practical experience of what would happen, thanks to an accident in a vacuum chamber in 1965 when a test subject's suit sprang a leak. The victim (who survived) stayed conscious for around fourteen seconds in a vacuum that wasn't as hard as that of space but still close enough to give the same effects. According to NASA, the exact survival limit isn't known but would probably be one to two minutes.

The extremes of clothing that enable us to explore the ocean or take a jaunt in space take us past the sort of evolutionary process that is ever likely to happen. Clothing has now, as a survival aid, taken human beings where evolution never could. (To be more precise, our upgrades have taken us where evolution is highly unlikely ever to. Good science rarely says "never." Living creatures have evolved to cope with the deep ocean and it is possible that some very basic forms of life could survive a trip through space, but it's incredibly unlikely that as complex an animal as a human being could ever evolve to cope with exposure to space.) In this sense, clothing hasn't just transformed us to Human 2.0 but to Human ∞.

Such clothing, however, is very specialized. Few human beings get to walk in space or even reach the bottom of the ocean. For most of us, clothes enable us to cope with environments where other mammals manage quite well, thank you, with their natural fur and reinforced feet alone. Yet for all the protective significance of clothing, it is by no means the only way we have upgraded ourselves to cope with the dangers posed by our environment.

Avoiding death was not a trivial task for early humans. They faced many threats, some of which were no less dangerous for being difficult to spot. The potential killers might be bacterial or viral, or could be food that contained chemical poisons. Over millions of years, some creatures have evolved a resistance to powerful natural poisons. Most birds, for example, are resistant to capsaicin, a potent alkaloid that blasts mammal nerve receptors to keep them away from a particular family of plants—it's the chemical that in tiny quantities gives chili peppers their punch. Rather than evolve a resistance to natural poisons, from early human times we were spreading the word—avoid certain plants. But the brain's real masterstroke to give humanity a boost above its evolutionary capabilities in this arena was cooking.

No one is sure just how cooked food became an important part of human life. It is generally assumed that cooking was first experienced accidentally, when an animal or grains were roasted by an accidental fire, perhaps started by a lightning strike. The attractive smell may have encouraged passersby to sample the char-grilled food, and the enhanced flavors would soon make it something that others would want to copy. Equally, it could have been that food was accidentally dropped into a fire or kept too near to a fire pit, as the human use of fire itself goes back a very long way.

Often portrayed as our first move into technology, humanity's relationship with fire seems to have started over a million years ago, before the final form of *Homo sapiens* had emerged. There is evidence of domesticated use of fire back to this period in the so-called Cradle of Humankind region in South Africa that appears to date back between 1 and 1.8 million years, and there are what are assumed to be hearths at Koobi Fora in Kenya that are around 1.6 million years old, though the precise nature of these African sites is the subject of considerable academic debate. Another fire hearth site on the Jordan River in Israel dates back around 790,000 years.

Whatever the numbers, we know that fire has been around a

long time as a survival aid. Initially its main use would have been for warmth and for protection. Fires helped keep predators away at night, as well as making low temperatures bearable. But the discovery of the effect of fire on foodstuffs should not be underestimated.

Initially, the attraction to cooking food may well have been the change in taste and texture produced by cooking. One of the effects of heating food is to modify the texture of proteins, making them easier to chew and digest. Cooking also releases complex chemicals that stimulate our sense of smell. We tend to think that when we eat food it is taste that drives our likes and dislikes, but smell is a very strong component of our system for detecting what is good to eat. You don't want to have to taste (for example) excrement to know it's not going to make a good meal. The sense of smell is the first line of defense, and much of what we think of as taste is actually smell. Some of the enhancement of taste in cooking is due to the breakdown of carbohydrates into simpler sugars and the concentration of flavors as water boils off, but much of it is due to the release of more aromatic chemicals to stimulate the nose and add to the apparent flavor.

Before long, though, a more important impact of cooking would have been noticed. It also has the advantage of killing bacteria and viruses and of destroying some toxins, such as phytohemagglutinin, the poison in kidney beans—which are deadly when uncooked—and the poisons found in nightshade-related plants such as potatoes. It must have taken some time to realize that not only was cooked meat tastier and easier to eat because it was less tough, but also those who ate it were less likely to suffer stomach pains and even sudden death. (The fact that this was because cooking killed microscopic bacteria would, of course, only be understood much later.) But once that realization was in place, humans had adapted to cope with food that in its natural state would have been inedible.

The realization must have been particularly difficult with a poisonous-when-raw food like the kidney bean. It's hard to imagine anyone seeing a neighbor die as a result of eating kidney beans who will then cook the beans and give them a try. It might have been the observation that cooking did make some inedible things edible that drove a hungry family to take the risk and thus to expand even more our self-adaptation to cope with otherwise dangerous foodstuffs.

Cooking, then, was a prime example of intervening to prevent something that could end life—a poison or a bacterium, for example. But the awareness of mortality would do more than inspire ways to protect ourselves from threats. It would also encourage attempts to lengthen our life span, initially by propitiating gods or invoking a magical influence that it was hoped would keep youth in place for longer. Nearly all primitive cultures ascribe life to a supernatural force, so it only seems reasonable that an appeal to the supernatural could lengthen life span.

The magical principle often invoked, whether consciously or not, was the assumption that life was in some sense a form of energy and it should be possible to extend life by taking this energy from others, effectively feeding on the "life force" to prolong life in the person employing the magic. This belief that something powerful could be evoked from life itself is the basis of all forms of sacrifice and led to some particularly unpleasant historical attempts to live longer than nature intended.

Most infamous of these excursions into magical extension of youth was that of the Countess of Báthory. Elizabeth Báthory (or Báthory Erzsébet in her original Hungarian) was born in 1560 and lived in Čachtice Castle (now in Slovakia). According to legend, to keep their youth Báthory and her co-conspirators bathed in the blood of young women. It's certainly true that they sadistically tortured and murdered many young women, and there is evidence that this was in part with the intention of extending youth, though the blood baths themselves may be fictional. Báthory was arrested in 1610 and died four years later without ever coming to trial.

The legends of using blood to extend youth would be mixed with the vampire legends to produce a heady combination that reinforced the idea of consuming the "life force" of others to extend life span. Although true vampirism is largely confined to legend, it seems unlikely, with such a strong belief in the nature of living creatures, that there weren't some attempts to make use of the blood or the death of others to extend life.

Much more certain, though, in the world outside of fiction was the alchemist's search for the philosopher's stone, the holy grail of alchemy. Alchemy, the predecessor to chemistry, dates back to several hundred years B.C. and combines the very practical and hands-on nature of physical chemistry with a spiritual component. This shouldn't be surprising—pretty well all early science sat comfortably alongside religious beliefs, and as late as Newton's time alchemy was considered a respectable attempt to search for the truths behind the nature of the world.

Although alchemy had a wider brief, throughout its history much effort was focused on the two goals that are still most frequently attached to it—the transmutation of base metals like lead into gold and the extension, perhaps to immortality, of human life. The mechanism for doing this is usually called the philosopher's stone in the West (a term that appeared in the original title of the first Harry Potter book, *Harry Potter and the Philosopher's Stone,* though it was changed in the U.S. edition to *Harry Potter and the Sorcerer's Stone,* as the publisher condescendingly thought the American public couldn't cope with the *P* word). As we shall see, this essential goal of alchemy was not necessarily literally a stone.

The tradition of alchemy goes back a long way, and it's difficult to pin down how it originated. There are three main sources, all with similar ideas and dating back well over two thousand years. The most direct influence on the Western tradition of alchemy was ancient Egypt, which in its turn would spread ideas to ancient Greece and hence, eventually, inspired the gradual shift from alchemy to chemistry that took place after the ideas of the alchemists arrived in Europe. It's certainly the Egyptian conception of alchemy that gave us the name, which comes from *al Kemia,* the old Arabic name for Egypt.

The ancient Egyptian traditions of the afterlife, including the practice of mummification, were very much tied up with an alchemical view of life. It was thought that it was only by the right preparation that a dead person could carry on his or her journey into a greater future life. In effect, the mummy was like the chrysalis from which a spiritual butterfly would emerge. This was not life extension in the sense of living longer in the normal way but rather leading someone on to a different, next stage of life.

There are strong parallels in ancient China, where the practice again was to try to preserve the body after death, as a chrysalis of the future being. A striking example of this was found in 1972 in a tomb at Mawangui in the Hunan Province of China. The tomb contained a body, identified as that of Lady Tai, who apparently died around 186 B.C. When the coffin was opened, the body is said to have been in perfect condition, as if only a few days after death. How the body was preserved is not clear—and it's not unusual for the level of preservation to be exaggerated in such cases—but this seems to have been an attempt to provide an extension of the earthly life.

Certainly Chinese alchemy aimed to produce some sort of elixir, sometimes called the pill of immortality, that would enable a human soul to fly out of its body and join the immortals. Chinese alchemy is inextricably tied into Taoism (pronounced "dowism"), the Chinese religion that has the concepts of extending life and immortality at its heart. Like other traditions of alchemy, the Chinese approach divided broadly into two searches—one for the "outer" elixir (*wai tan*), which was a physical route to preserving life using potions of herbs and minerals, and another for the "inner" elixir (*mei tan*), which relied on a spiritual approach, using methods like prayer, meditation, breathing techniques, sexual practices, and exercise.

It is the outer elixir that corresponds to the philosopher's stone, and in the Chinese tradition the end goal was always the extension of life, so any talk of changing lead into gold was less about metalwork than it was symbolic of transforming the "lead" of ordinary life into the bright, untarnishing, eternal metal that was the gold of the undying life.

The third of the major sources of alchemy was India, where the approach had many similarities with the Taoist tradition. It has been suggested because of the parallels between the different traditions that all three of the ancient civilizations that espoused alchemy got their ideas initially from a common source, though there is no direct evidence for the existence of this proto-alchemy.

Whichever tradition you examine, the approach is much the same. Ingredients are processed in a lengthy way, often repeatedly heating and cooling them, distilling and reacting the mixture to produce the ultimate goal. Because alchemy doesn't separate the physical and the spiritual, there was often a requirement to perform the processes at certain phases of the Moon or dependent on the position of planets in the Zodiac, and the alchemist had to be in just the right frame of mind, properly prepared for what he or she would undertake.

This meant that there were many, many things to go wrong, so it seemed quite reasonable than when anyone tried out a reaction in the alchemical tradition it didn't work (though everyone, of course, knew there were people who claimed to have succeeded; or at least, they knew someone who had heard of someone who had succeeded). When Peter Marshall, a researcher into the history of ideas and a confessed enthusiast for the romance of alchemy, was trawling the world searching for modern-day practitioners of the art he received a very revealing answer to a question he posed while in India. "Do you know any recipes [for elixirs]?" he asked. He was told ". . . they are never the complete story. There are still some secrets not said; there's always something missing." The constant failure attributed to "something missing" could equally, and perhaps more reasonably, be interpreted as an illustration of the fact that there was no elixir, there was nothing to work—the dreams of alchemy producing an elixir to extend human life are fantasy—but human nature is reluctant to accept such a message.

Whichever tradition first gave birth to the idea of the philosopher's stone, the miraculous substance that could provide eternal life, it was never intended that this should be taken literally as a simple piece of rock. This "elixir of life" (a term that now suggests a fluid) usually seems to have been a powder, involving materials that had been transformed through a process that made them first black, then white, and finally red. Each color had a symbolic status in the gradual purification of this mythical substance. One obvious ingredient, because of its beauty, rarity, and permanence, was gold, but worryingly the most commonly used material was mercury, perhaps because of the living, flowing quality of the only metal that is liquid at room temperature. Unfortunately, mercury is a deadly poison and many alchemists seem to have succumbed to the vapors as they heated vials with a mix of mercury and other ingredients or consumed the deadly mixtures they produced.

It has been suggested that Isaac Newton, who investigated alchemy for many years, was suffering from mercury poisoning when his behavior went through a sudden, violent change. During a few months in 1693, Newton seemed to tip over the edge into madness. Although he was often an unpleasant character, Newton usually maintained a cold logic to his communications, but in this period he wrote to his friend the diarist Samuel Pepys, then president of the Royal Society (Britain's equivalent of the National Academy of Sciences), to say that he could never see him again, because of some unspecified slight. Soon after, Newton accused philosopher John Locke of attempting to embroil him with women. This atypical behavior could well have been caused by inhaling mercury vapor.

It's ironic that in trying to extend human life many alchemists must have succeeded in making people sick or even managed to kill them. However, not all the substances described as mercury in alchemy were the real thing. Just as the idea of converting base metals to gold was often a symbolic process that implied the purification of a material, so the ingredient referred to as "mercury" was probably often merely the component in the mix that was thought to have the *character* of mercury. This is typical of the mystical indirectness used by alchemists that makes researching the subject seem like trying to catch a squirt of gas in a fishing net.

In some cases the smoke and mirrors were clearly an attempt to defraud. There were plenty of supposed alchemists who were nothing more than hustlers hoping to make a quick buck. They were the snake oil salesmen of their day, out to extract money from innocent people who had a longing to extend their life, a trade that had much in common with the quack medicine peddlers common in the nineteenth century. Only the patter changed—what was being sold, an unpleasant and sometimes dangerous mix of herbs and minerals, stayed pretty much the same (and can still appear under the banner of "alternative medicine" alongside more respectable cures today). But not all alchemists were fraudsters. Many believed in what they told their public, and for them, the evasive approach was taken just to keep their material secret from the common herd.

A lot can be gleaned about the attitude to secrecy—and about the approach taken by alchemists in the attempt to extend human life span—from a remarkable document written by the medieval friar Roger Bacon. Called "Concerning the Marvelous Power of Art and Nature, and Concerning the Nullity of Magic," the long letter, probably written around 1250, sets out to show that what many assume to be the result of magic is in fact purely a combination of natural science and art—by which ("art") Bacon meant the work of human hand, the produce of an artisan.

Bacon told his readers that it was not wise to write secrets openly and provided classical quotes to justify this, like "[t]he man who divulges mysteries diminishes the majesty of things, and a secret loses its value if the common crowd knows about it." Alchemists, ever enthusiastic to keep their secrets to the special few, were in the habit of writing down their formulations for the means to preserve and extend life in obscure terms, giving this indirect formulation a title like "the philosopher's stone" as a way of obscuring it. It's interesting that when later his letter Bacon describes what he calls "the philosopher's egg," he is, in fact, giving a recipe for making gunpowder.

Most relevant, though, for an insight into the alchemists' attitude to long life is a section of his letter where Bacon describes what he calls "the ultimate attainment, in which the whole complement of Art joined with all the power of Nature is effective"— the prolongation of life. Bacon lists what he believed to be examples of the historical reality of the philosopher's stone. He tells of a man whom the Roman philosopher Pliny (technically Pliny the Elder) describes. This man "lived in his probity beyond the accustomed age of man" as a result of "using oil externally and mulsum internally." (The "mulsum" Pliny refers to is thought to be a mixture of honey and water.)

Other examples Bacon gives are a case attested to by the evidence of a papal letter of a man called Almanicus, who while a captive of the Saracens took a medicine that apparently extended his life to five hundred years, and we are also told of the seemingly widely known knowledge that farmers "frequently attain the age of a hundred and sixty years or thereabouts," apparently as a result of their knowledge of "plants and stones." Few farmers I know have managed to reach this age—unfortunately, although

Bacon was arguably the first scientist, he wasn't very good at science (not surprising if he was the first) and had a habit of equating hearsay with scientific evidence.

A lot of the logic for the practicality of extending life in the way that Bacon describes comes from the ages given for some of the early characters in the Bible. Adam, we are told in the Bible, lived 930 years, Methuselah 969 years, and others of the same period had life spans in the region of 800 to 900. As Bacon puts it, "after the fall, a man might live for a thousand years; and since that time the length of life has been gradually shortened. Therefore it follows that this shortening is accidental and may be remedied wholly or in part." Bacon, with his early scientific view, had a surprisingly modern idea of why we no longer attained these biblical ages. Short lives were not, he thought, "from the heavens" or a result of sinning but were due to defects in the way people took care of their health.

We still tend to hold on to another number from the Bible—"three score years and ten" or seventy years—as an idea of the natural length of life, but what has human life expectancy really been like through history? With what was Bacon comparing the lengthy lives of the Old Testament characters? Average life expectancy has grown phenomenally in the last hundred years. From the dawn of history through to the nineteenth century, average life expectancy has been between twenty-five and thirty-five. Now it is in the high sixties, and higher still outside Third World countries.

How, then, did the writers of the Bible come up with the apparently overinflated but visionary figure of seventy? This is because the average figures are very misleading. The historical figures are dragged down immensely by a very high infant mortality rate.

Before modern medicine, most children would not make it to adulthood. Similarly, many women died while giving birth, in their twenties or younger. If early deaths are excluded from the average, life spans in the fifties, sixties, and seventies were not uncommon. (Roger Bacon is thought to have made it into his seventies.)

Typical life spans of those who survived into adulthood dropped as we moved from the preindustrial to the early industrial age, though more children were surviving, so this isn't clearly reflected in the averages. Now childhood mortality is at an all-time low. As Armand Marie Leroi points out, 1994 was a remarkable year in this respect. In 1994 no eight-year-old girls died in Sweden—not a single one. While this was just one point in the statistics—the next year, no doubt a handful did—it is still a notable fact that would have been inconceivable to our medieval ancestors. When there's a funeral for a baby or an older child it is always a very emotional and particularly sad occasion—it's sobering to think that not many years ago, and throughout all of history before that, the majority of funerals were for babies and older children.

Going back to the Bible, there is no scientific evidence for the reality of the extreme biblical ages of the patriarchs, but uncannily, the age spans ascribed to the biblical elders have been predicted as possible for the future by at least some modern gerontologists (scientists who study aging). The most radical of the fold, Aubrey de Grey of the University of Cambridge, has said: "I think the first person to live to 1,000 might be 60 already."

Startlingly, there are some who would go even further than this. Ray Kurzweil, a millionaire pioneer in IT areas such as optical character recognition, now concentrates more on life extension than information technology and has every intention of living forever: not just having a long life but *never* dying, unless it's by accident. That's quite a plan. Kurzweil's argument is intriguing, though few gerontologists would go all the way in agreeing with him. He sees three steps to indefinitely prolonged life, steps that he believes could be available to anyone who is likely to live another twenty years on current estimates—which just about takes in the baby boomer generation of which Kurzweil is a part. "Immortality," says Kurzweil, "is in our grasp."

The first of the three steps, or bridges as Kurzweil calls them, is to do as much as we can with the best current knowledge of what it is that ages and kills us. The aim is to prevent the onset of aging as much as possible to be able to hold on until the next technology bridge (of which more in a moment) is available. Kurzweil and collaborator Dr. Terry Grossman have a program that takes in everything from avoiding sugar and drinking green tea to following a serious regimen of dietary supplements and exercise.

Many of the duo's recommendations taken individually are familiar from government health campaigns or the ideas of alternative health gurus, but in almost every case Kurzweil goes further, determined to squeeze the last drop out of the possibilities for extending life. For example, where most government campaigns suggest we eat lots of fruit and vegetables, Kurzweil recommends only moderate fruit consumption (but loads of vegetables), to reduce the sugar consumption that goes hand in hand with eating fruit. (Sugar can certainly creep up on you. Even apparently harmless skim milk has a surprisingly high sugar content.)

Part of the need for government health campaigns, or more dramatic modifications of what we consume, is a direct reflection of the way we have made changes to our natural capabilities. In the past, inadequate diets tended to kill people off relatively early. As we have changed the way we eat, we have not only compensated for this but also gone too far the other way. Nutrients like fat and sugar that were in relatively short supply, giving us a genetic tendency to grab them and savor them whenever we can, are now more than plentiful. The result is a diet that has diverged from our evolutionary norm, with much too much fat (particularly saturated fats and trans-fats), too much of carbohydrates with a high glycemic load (these are carbohydrates that have more tendency to produce sugars in the body) like bread, pasta, and rice, and too much refined sugar.

To see just how far it's possible to go in compensating for our movement away from an ideal intake, you only have to take note of Ray Kurzweil's own routine. As well as following a strict diet and getting plenty of exercise, he takes around 250 supplement pills a day (that's an amazing 91,000 a year) and makes a weekly visit to a clinic where he spends a whole day undergoing various therapies, including six intravenous nutritional supplements and acupuncture. It's unlikely that many others will go as far—but Kurzweil is past his midfifties and so wants to maximize his chances of surviving long enough to reach the second bridge.

Some of the supplements Kurzweil suggests are considered at best borderline by most medical experts, though it should be stressed he hasn't jumped on every bandwagon. For example, human growth hormone (HGH) has been widely touted as a mechanism for reversing some of the effects of aging after a study in 1990 found a move of mass from fat to muscle and better levels of cholesterol, insulin sensitivity, and bone density. However, HGH has only been properly trialed for its intended use, helping those with growth difficulties, requires an intensive treatment, and has the potential for significant dangerous side effects, so it is probably better to make use of lifestyle changes (for example, cutting down sugar and increasing sleep time) that have the effect of stimulating natural growth hormone production.

The second bridge that Kurzweil believes will take over from eating well and exercising involves what he refers to as the biotechnology revolution. This is the knowledge we are gaining, particularly from genetics, to be able to actively switch off features that promote age and death and to turn on aspects of our genetic code that enable youth and energy. This second bridge, Kurzweil believes, will keep us going long enough to reach the third stage, which he envisages being significantly possible by the 2030s. Traditionally the stuff of science fiction, this third bridge involves the ability to rebuild and improve on functions of the body and brain using nanotechnology and artificial intelligence. If that bridge is ever crossed, then it will be a significant addition to the upgrades we have seen in human capability so far—perhaps a little like jumping from Windows 1.0 to XP in one go.

Even Kurzweil admits that it could be as late as the end of the twenty-first century before his third bridge is completely viable, but he points out that through detailed mathematical modeling of technology trends he has found that we depend too much on looking back at the past to predict the future. He compares the one hundred years of development in the twentieth century with about twenty years of change at present, because our rate of change is so much faster. Technology is not just getting better; the rate at which it gets better is accelerating. We should, he believes, make as much change again in the next fourteen years as in the last twenty, then the same amount again in the following seven.

If true, this remarkable growth rate would mean that the twenty-first century will see the equivalent of twenty thousand years of progress at current rates. It's certainly true that it doesn't make sense to simply compare the changes in twenty years in the past with the changes in twenty years now. (If you have any doubts, compare the technological developments between 1350 and 1370 with those between 1950 and 1970.) A useful yardstick to measure the rate of development is the change in our ability to sequence genomes, unraveling the details of a living thing's DNA. In the late twentieth century it took fifteen years for the HIV genome to be sequenced. When the respiratory disease SARS (Severe Acute Respiratory Syndrome) broke out at the end of 2002, the genome of this comparably complex virus took thirty-one days to sequence. The U.S. National Science Foundation has predicted that the key technologies for life extension—biotechnology and nanotechnology—will grow exponentially for decades.

Before I examine some of the implications of Kurzweil's final bridge, which will certainly come at some point, it is worth examining his overall thesis—that because of the explosion in understanding of the body's mechanism, the advances in genetic technology, and the rapid steps being taken in nanotechnology, there are already those alive—and Kurzweil inevitably includes himself—who will, effectively, live forever.

There is certainly one appealing piece of logic behind this. Kurzweil points out that many experts believe that by the mid-2010s we will be adding more than a year to life expectancy with every twelve months that passes. (It's relatively hard to find these "many experts," however.) If we can survive that long, in principle we can ride the wave, picking up on each new discovery as it comes along. It's certainly an appealing picture.

What is notable, however, is how much of a resemblance there is between Kurzweil's grand idea and the strivings of alchemy. Underlying alchemy is a considerable degree of self-deception. Very few of us want to die, so we will clutch onto anything that we think is going to extend life. After all, until you do pass away (and at the point you won't know any better) the solution you choose, whatever it is, appears to have worked. I could say that eating ten Big Macs every day is a way to live to a hundred. I don't recommend it, because in practice it will probably shorten your life, but until a big enough sample of people have kept up this diet and have died, it can't be shown to be untrue (just very, very unlikely).

As yet it's much too soon to say whether Kurzweil's regime will really work or if he has suffered all this manipulation and swallowed so many pills (not to mention enduring what sounds to be a very dull diet) for nothing. We do know, for example, that UCLA medical school professor Roy Walford, who advocated a very low–calorie diet and followed his unpleasant regime himself with the aim of living well past one hundred, died at the age of seventy-nine in 2004. It's a little over the average for a U.S. male but not enough evidence (you would need many examples) to prove anything—and not many of us would want to stick with Walford's dietary intake of sixteen hundred calories a day to add a year or two to our life expectancy. It's certainly true that we now know much more about what various dietary components do to our bodies than was known in the past and Kurzweil's first bridge is based on the best scientific knowledge rather than the pure guesswork of the alchemists—but equally there is no certainty.

Realistically, Kurzweil's models of future development rates are going to be flawed. It's the nature of future gazing. We can dress it up by calling it prediction or futurology, but the fact is, what we're dealing with is guesswork. No one can predict developments accurately in as relatively simple an environment as the stock market (nor can they give a guaranteed prediction of the next Kentucky Derby winner). To attempt to predict the rate of development in any technology can amount to hubris, as many a maker of a science fiction movie has found to their cost. If you doubt this, watch *2001: A Space Odyssey,* acclaimed at the time it was made for the accuracy of its science predictions. Not only did it portray a manned spaceship to Jupiter launched at a time when we hadn't taken men farther than the Moon; it also showed commercial space shuttles plying between a space station and the Earth, operated by Pan Am. Remember Pan Am?

However, Kurzweil could be right. Aubrey de Grey of the University of Cambridge has said that we should be able to successfully stop aging in mice within about ten years. This might not seem too exciting (except for mice), but we share a high percentage of our genetic code with the mouse, and he believes that this will make it possible to halt and even reverse human aging— the second of Kurzweil's bridges—five to ten years after that.

It might be easy to assume that de Grey, who is not a conventionally trained gerontologist with a biology background but was originally a computer scientist, is just throwing out idle speculation, but his work, with respected researchers in the field, including Bruce Ames of the University of California at Berkeley,

Andrzej Bartke of the Southern Illinois University School of Medicine, and Judith Campisi of the Lawrence Berkeley National Laboratory, suggests that there are specific, positive actions that can be taken to stop and reverse aging, a process de Grey refers to as Strategies for Engineered Negligible Senescence, or SENS.

According to de Grey there are seven key problems of aging that need to be addressed, problems he refers to as the "seven deadly things." (Think "sins" with a lisp.) For each of the deadly things, he and his colleagues suggest a specific set of solutions. Some are straightforward and easy to do today (exercise, for instance). Others require the development of techniques for manipulating genes and other aspects of the body at the cellular level that are at least several years away and are known at the moment to have all sorts of potential side effects that might be hazardous or even fatal.

The first of the deadly things that come upon us as we age is cells that are lost entirely or that wither away, a problem that is particularly relevant to the brain in old age and to the diseased heart. Cell loss and atrophy can be countered partly naturally by exercise, whether it's physical exercise for muscles such as the heart or mental exercise for the brain. Cells can also be encouraged to divide, replacing loss, by using signaling chemicals called growth factors, while new cells can be introduced to a failing area with the help of stem cell therapy (see page 43). Dealing with replacing cells is one of those areas where extreme caution has to be invoked before we go beyond "natural" solutions like exercise, as many methods of encouraging cell division can also increase the chances of an outbreak of cancer, which is, in effect, uncontrolled cell division.

The second of de Grey's answers to aging deals with nuclear mutations—that's "nuclear" as in the nucleus of the cell, mutations to our DNA, not "nuclear" as in "nuclear power." De Grey claims that on the whole our inbuilt mechanisms can cope with these natural mutations, except in the case of cancer, where just a single cell with mutated DNA can kill us. Cancers overcome the natural "off switch" our cells are fitted with, the telomere, a section at the end of a chromosome that has a repeated pattern, rather like the numbered paper tickets you are sometimes issued with when waiting in line.

When a cell is duplicated, one of the telomere tickets is "pulled off"—when they run out, the cell can no longer duplicate. This mechanism stops a cell from dividing too many times. Cancers activate telomerase, the enzyme that is used when a new human being is created, to reset the markers that limit cell duplication. De Grey suggests that by eliminating the genes that produce telomerase, along with genes that produce another enzyme with a similar effect, we could also eliminate cancer.

There are drugs being developed to suppress telomerase production, but de Grey's suggestion is a more complete, and more drastic, solution. This is not an easy option (but then, curing cancer across the board is a big prize). Not only would you have to be able to target the right genes and knock them out; it would also be necessary to replace stem cells like those in the blood, special cells that are intended to divide on a regular basis and that would otherwise have relied on telomerase. Cells like this would have to be replaced about once every ten years to keep the body functioning. By current standards, any life extension treatment that requires subsequent regular maintenance to stay alive is extreme, to say the least. Yet the goal of this particular part of SENS is awesome in scope.

Third on de Grey's hit list is a peculiarity of the way we, and all other complex cellular-structured living things, are put together that dates back billions of years—tiny components of the cell called mitochondria. These minuscule pods floating within the cell are sometimes called its power plants, as their job is to take the oxygen from breathing and combine it with chemicals from our food to make ATP (adenosine triphosphate), the molecule that the body uses to store up energy. It is thought that mitochondria were once bacteria that became part of the cell in a mutually beneficial symbiosis.

Mitochondria have their own DNA—just thirteen genes that are separate from our main chromosomes in the nucleus of the cell. Unlike our principal body of DNA, which comes from both our parents, the mitochondrial DNA only comes from our mothers. The genes in our mitochondria are nowhere near as well protected against mutation as our main set of genes and are particularly in danger from attack by free radicals (see page 59), so de Grey suggests a dramatic solution. Move the mitochondria's DNA out to the nucleus, shifting those thirteen genes into the hands of our main DNA management process.

This isn't as mad as sounds, because the mitochondria already have a mechanism for using genes from the nucleus. These built-in ex-bacteria need the action of around one thousand genes to work, and all but the thirteen are already sited outside of them in the chromosomes. They have migrated there over time. All de Grey is proposing is finishing the job that has gradually moved mitochondria from being independent bacteria to a totally dependent part of the cell mechanism. By adding these thirteen genes to the genome, suitably modified so the messengers that enable the existing internal genes for the mitochondria to communicate will also work with them, the risk of mutation in the mitochondrial genes will be much reduced.

This has already been done in 1986 at Monash University, admittedly only with a single gene in a yeast, but the concept is at least not totally unproven. Like many of the solutions to the seven deadly things, this approach is not without dangers, even if we could make it happen today. The cells that suffer worst from mutations to the mitochondrial DNA are those like brain cells that don't divide, so we need to use gene therapy (see page 47) to get the changes into the cells—but de Grey emphasizes how quickly gene therapy is advancing.

Number four on de Grey's hit list is almost the opposite of the first. The first deadly thing was cells we need that instead die off. The fourth deadly thing is cells that overstay their welcome or weren't really wanted in the first place. Fat cells tend to replace muscle over time, an undesirable progression, while some extra, malfunctioning cells called senescent cells become more common as we get older, and we can also develop an excess of immune cells like our white blood cells. Some of these cells that have the job of fighting infection become too numerous over time or stop working. Compared to some of the problems, this is a relatively well-understood one, and the mechanisms to encourage the unwanted cells to die off or to allow the immune system to destroy them are relatively simple, even though we don't have an immediate solution available at the moment.

The fifth deadly thing is a tendency for some long-lived proteins in the body to undergo gradual chemical modification. We see this in the hardening of the arteries and the loss of transparency in the lens of the eye. Here two proteins are linking together, resulting in a more rigid chemical structure. A drug that lowers blood pressure by breaking these chemical bonds is already under testing—it only deals with one particular protein, but this establishes a model that could be used to attack this issue.

The sixth deadly thing that de Grey identifies is "extracellular junk," chemicals and small structures outside of the cell wall that have no constructive function and can cause problems. An example of this is amyloid, the chemical that forms into masses called plaques in the brains of sufferers of Alzheimer's disease, and there are several other similar problem bits of junk around our bodies. Work is under way on a vaccine to encourage the immune system to take on the junk it currently ignores—alternatively, other chemicals could be developed that actively break up plaques.

Last, and seventh, of the deadly things is junk within the cell. Here the garbage disposal mechanism of the cell can take on some large structures that it can't break down. In cells that don't divide, like brain cells, this rubbish can gradually build up until the cell clogs up. Atherosclerosis, plaques on the artery wall that break away to cause heart attacks and strokes, is caused by this kind of poor cellular junk management. De Grey suggests giving cells extra enzymes to help break down this junk, probably extracted from soil microorganisms, which make a living out of junk disposal. His group at the University of Cambridge is now working to find such enzymes.

To address de Grey's seven deadly things or to put Kurzweil's second bridge in place (which goes even further) we needs a mind shift in medicine, a shift that does seem to have started to take place. Traditionally medicine has been about fixing what is broken. We wait until something goes wrong, then repair it if we can. In historical times there was little choice about this. Medicine, as we will see in chapter 6, grew up in a particularly unscientific way. As there was no understanding of how the body worked, medicine could only ever be a trial-and-error attempt to counter things that had already gone wrong.

Now our position is radically different. Increasingly, drugs and operative treatments are based on an understanding of how the body works down to the cellular, genetic, and biochemical level. And with that understanding comes an opportunity not only to treat disease but also to take positive action to enhance the body, protecting it against future failings. It is essential if Kurzweil's dream is to be fulfilled that we move from reacting to problems to using our understanding of the body's mechanisms, taking precautionary action to prevent things from going wrong in the first place.

A lot of Kurzweil's second bridge depends on our growing understanding of genetics, both dealing with disease by actively manipulating genes and in realizing that each individual has particular restrictions on his or her ability to live longer that are determined by his or her genes. As it becomes easier to map a human's genome (the collection of genes for a particular individual) and to understand which of an individual's genes are unique, it should be possible to structure our treatment to match the specific requirements that this one person has, thanks to his or her unique genetic makeup.

Our genes have been described as the blueprint for making a human being, though this is a flawed analogy. Every one of our cells contains around twenty-four thousand genes. Each of these genes is a set of coding instructions, stored along the length of the very long molecule DNA. The job of a gene is to provide instructions on how to build particular proteins, the complex molecules that act as agents and machines to make things happen throughout our bodies. It's because of this mechanism that genes really aren't blueprints. A detailed blueprint of something as complex as the human body would take much more information than is held in our genetic "library," the genome. Instead, it's more like a recipe.

To make the distinction clearer, imagine the difference between a "blueprint" for a cake and a recipe for the same dessert. The detailed blueprint would have to say exactly what molecular constituents are placed where in each and every tiny crumb of the cake. The recipe simply says mix A, B, and C, then cook for thirty minutes at 350 degrees. The genetic instructions have more in common with the recipe (though the body is, of course, much more complex than a cake) than they do with the blueprint.

Initially, with today's technology, there are limits to our ability to manipulate genes. We can't just reach in and fix a faulty gene, correcting a genetic disease or removing a factor that could contribute to an early death. However, we can already influence gene expression. Expression is the means by which genes make things happen. Of themselves, genes are effectively just blocks of information. Gene expression involves reading that information and translating it into action, through the construction of functional proteins, the complex chemicals that effectively run the body.

While we can't easily manipulate the genes themselves, we now have plenty of medicines that can influence the way gene expression occurs—as can the many natural supplements that Kurzweil advocates. The first part of Kurzweil's bridge two is to know your own genetic makeup and from that to tailor the treatments that are used to keep you alive longer. A full scan of your genetic makeup would currently be extremely expensive and pretty well impossible to interpret. One of the developments that is essential over the next few years as this technology progresses is making the analysis of all the information present in a single genome more practical. For the moment, though, it is possible to look out for specific faults that are known to present risk factors for extended life.

Most of the human genome—the total information held in our genes—is the same from person to person, but the small percentage that makes each individual unique even if raised in the same environment consists of small variations in the individual "letters" that make up the code of our DNA. We each have hundreds of thousands of these variants, referred to as SNPs (single nucleotide polymorphisms). Not only do they make us unique; they can also make an individual more or less susceptible to the breakdowns of the body that could cause a life to end. By checking for the most common known risk SNPs it is possible already to give some guidance on the appropriate treatment required to provide some life extension.

Similarly, our increasing understanding of the genome has made it possible to produce "designer babies," engineered to avoid certain defects. Currently this involves selection, rather than modification, in a process known as preimplantation genetic diagnosis, which was first used in 1990. A single cell is removed from an embryo when it is only a few cells in size, and the genes within are checked for known genetic faults. The embryo is only implanted by in vitro fertilization (IVF—test tube babies) if the check comes up safe. This technique has been used successfully to select from embryos at risk of genetic diseases like Huntington's and cystic fibrosis.

A more dramatic example was baby James Whitaker, born in June 2003 in Sheffield, England. He was selected not to avoid a particular problem but because his embryo had a particular characteristic. James's older brother, Charlie, suffered from an unusual type of anemia that could only be helped with an injection of stem cells (see page 43) that were appropriately tissue matched. James's embryo was selected as the one most suited to provide a tissue match for his brother. Stem cells were then removed from the umbilical cord when James was born, to be used in the future to help Charlie.

Ian Wilmut, one of those responsible for cloning Dolly the sheep (see page 64), believes that there could be a way both to correct genetic faults and to make IVF more reliable. Wilmut is strongly against reproductive human cloning, the process of making a cloned individual from another human being. He argues against this both practically—all the evidence is that most clones would die or be very unhealthy—and ethically. However, Wilmut believes that the approach used to clone Dolly, where the DNA of an "emptied" egg is replaced, a procedure called nuclear transfer, does hold out hope for new developments in a few years' time.

At the moment, many IVF embryos fail to implant, and if there is an embryo with a genetic fault we can't fix it; we can only select out the ones that seem better. Wilmut suggests it will soon be possible to attempt fixes on stem cells from an embryo, check if they worked, then use those stem cells to provide the DNA that goes into the "empty" egg. The big advantage over traditional cloning is that the new embryo will be started with original stem cells, not from adult cells. This means there will be fewer problems, and also that the baby produced will not be a clone of another human being, just a clone of a cell that never became a human being—the baby will be a unique individual, not a copy.

This is not a piece of work to be undertaken lightly. As Wilmut acknowledges, there is a big difference between making a genetic change to a stem cell that will form an embryo and gene therapy in the way it has been applied to date. The current type of gene therapy, called somatic therapy, only influences a section of the cells in the body. However, making a change to the original cell from which the embryo forms will result in a modification to every cell in the body. The master DNA is being modified. Such a change, called germ line therapy, causes significant concerns among biologists, as it has the potential to spread a genetic modification through the human gene pool. The change may be passed on to any children and gradually become the norm. It's the same concern that many feel about genetically modified crops—that they could "escape" and the changed genes propagate through the species until there is no untainted version.

If we could be sure that a change was purely beneficial, there wouldn't be a problem. No one would worry too much if the genetic change that prevented cystic fibrosis, for example, escaped beyond the person being treated and meant that no one ever got cystic fibrosis again. But in practice, we know from the experience of existing attempts at gene therapy that modifying a gene can have unexpected consequences—and if those proved dangerous, there is a real fear of what would happen if this modification escaped and spread.

Returning to Kurzweil's second bridge, the simplest component involves modifying the environment in which gene expression takes place, influencing the outcome. The next possibility is to take a more direct line of attack and either block or modify the way the risky genes express themselves. Although it's mostly at the laboratory level, there is now significant work under way to control this expression process, so it should increasingly be possible to switch off hazardous genes and to switch on those that tend to extend life.

The final part of Kurzweil's second bridge involves getting our hands on the genes themselves. There are mechanisms to do this. A virus, for example, is a natural biological device for manipulating genes, and it's possible to take the "attack" mechanism of a virus and replace its payload with a more desirable structure. As yet, though, such gene therapy is crude and the outcomes difficult to control. It's a bit like throwing a single word at a huge magnetic whiteboard that is covered with the entire text of a novel and hoping it will land at the right place. Some early attempts at gene therapy have resulted in deaths when the virus mechanism itself caused problems or the genetic material went into the wrong part of the patient's DNA. These problems are difficult but will be overcome, and there is no doubt that gene therapy holds out huge potential benefits for extending life span, even though it has sometimes been oversold as being nearer consistent safe use than it truly is.

Although there have been failures, there have already been successes from the crude gene therapy that is possible so far. One of the first people known to benefit was Ashanti DeSilva, a five-year-old girl treated at the NIH Clinical Center in Bethesda, Maryland. Each of Ashanti's parents had a fault in one of their genes, a gene that is responsible for the production of ADA, a protein that helps keep our immune system going. Ashanti inherited both the faulty genes, leaving her without the ability to produce ADA. Without this gene in action, Ashanti had to take heavy medication in the form of injections of an enzyme similar to that produced by ADA to stay alive. She was a "bubble baby" who had to be isolated from all possible infections if she was to survive.

The team at Bethesda used a virus to insert a working version of the gene into some of Ashanti's blood cells, then reinjected the modified blood back into the girl. The result was a startlingly impressive recovery. Ashanti still needs some medication but nothing like what she had to take before. Gene therapy has indubitably worked here, but it's important not to underplay the difficulties and risks. Often a problem is caused by multiple genes, so it is not enough to fix one gene, yet at the same time one gene can have multiple roles, so an attempt at fixing a problem could also wreck one of the gene's positive functions. Not only can modified genes be delivered to the wrong place in the chromosome; modifications can have unexpected side effects also. What's more, there are so many defective genes left in the body that haven't been modified (remember there's a set in every cell) that there's usually a need to repeat the therapy many times to keep the results working.

Gene therapy like Ashanti's holds out huge hope—and in more than a decade since her first therapy the techniques have gotten much better—but there are still huge problems with working this way. As recently as 2007, Ashanti was still the only person to receive successful gene therapy for her particular problem, severe combined immunodeficiency—all other attempts have failed. In one example in 2000, two chidren died in an attempt to cure the same condition when the virus used to carry the modified gene inserted itself into the wrong place and triggered a form of cancer. However, thousands of trials have now been undertaken in other fields, and in some areas, notably cardiovascular disease, HIV, and some forms of cancer, there have been positive results.

Scientists who work on preimplantation genetic diagnosis are now looking to take the major step of moving to true genetic engineering—not just selecting the most favorable embryo but also repairing defective genes or, in principle, making modifications to encourage certain traits in the child who will be born. There is much dispute over just how far this can be taken. Most would agree that it is desirable to make modifications to remove known genetic faults that would otherwise blight the child's life, but few would argue that a preference for (say) a child with blue eyes is one that should be pandered to (see page 132 for a further discussion of this ethical dilemma). In terms of the essential message of this chapter, though—extending life—there is no doubt that by debugging the DNA of fetuses we have the potential to extend what would otherwise be shortened lives.

The third bridge in Kurzweil's vision, the incorporation of nanotechnology and AI into the body, is one where Kurzweil believes that we will become Human 2.0, a new evolved being. The thesis of this book is that we are already well past the 2.0 stage, but Kurzweil is only thinking biologically, and there is no doubt that his bold vision would take us much further down the adapted evolutionary track. There are already projects under way constructing tiny mechanisms to be injected into the bloodstream to monitor conditions or to deliver tiny doses of hormones or other chemicals into the body. It is inconceivable that this research will not lead to more and more tiny devices with an increasing level of intelligence that operate within the body to keep us functioning for longer.

Among the possibilities Kurzweil suggests are on the cards are self-propelling robotic replacements for blood cells (this eliminates the importance of the heart as a pump, failure of which is one of the most common blockages to living longer), built-in monitors for any sign of the body drifting away from ideal operation, nano-scale robots that can deliver drugs to control cancer or remove cancer cells, and even robots that make direct repairs to genes. (By "nano-scale" I am referring to devices around a nanometer in size, making them comparable in size or smaller than the "molecular machines" that drive our living cells. A nanometer is one billionth of a meter. To give a better idea of the scale, the point of a pin is around 1 million nanometers across.)

In a rather strange development, Kurzweil also expects we might separate the pleasure of eating from getting the nutrients we need, leaving the latter to nano-scale robots (nanobots) in the bloodstream that would release what we need, when we need, while other nano devices would remove toxins from the blood and destroy unwanted food without it ever influencing our metabolism. You could pig out on anything you wanted, all day and every day, and never suffer the consequences.

Is Kurzweil's vision sober reality or just a pipe dream? Research into extending life has been an academic pursuit since the early days of the twentieth century, but it doesn't have a great pedigree. Michael R. Rose, an evolutionary biologist from the University of California at Irvine who has made a career out of researching aging in fruit flies, comments that when he first got into the field in 1976 this was not a hot topic. As he puts it,

"There was no bandwagon, not even a minivan." This lack of enthusiasm (Rose himself had no interest in the subject and only got into it as a way of getting close to a scientific hero of his) was largely because the early attempts to understand aging had ended in failure.

Early suggestions for the causes of aging ranged from the toxic effect of bacteria in the gut (not entirely far-fetched as something that needs controlling but proved not to be the cause of aging) to the damage caused by the gradual accumulation of mutations in our DNA. None of the theories proved effective as an overall explanation of why aging occurred, though as Aubrey de Grey has shown (see page 43), DNA mutation is a contributory factor to dying. Working with the tiny fruit flies *Drosophila,* Rose eventually became fascinated by the challenge and found that he was able to breed what he referred to as Methuselah flies, insects with an extended life span. His work proved very successful, though he confesses he never even gave any consideration to the thought that this might be relevant to human longevity until the 1980s— when he himself was reaching the sort of age where mortality begins to remind us we aren't going to live forever.

Of itself, knowing that you can breed fruit flies for longevity might not seem a great help. The idea certainly wasn't to revisit Báthory-style pseudoscience and inject extract of Methuselah fly, or bathe in the fluid from crushed flies, to try to live longer. The important thing, though, is that with our increased understanding of biology at the genetic level it is now possible to discover the differences between the long-living flies and their more transient relations and to use that knowledge to devise means by which we can extend our own lives.

When Rose first realized this (and became enthused with the

possibilities of living longer), he assumed he would have to move on from flies to something closer to humans—in all probability, the long-suffering laboratory mouse. He even took a step in that direction, attempting to set up a company in the heady early days of biotechnology spin-offs that would produce and understand Methuselah mice, a much longer project than working with the flies, as mice don't reproduce anywhere near as quickly. But as it became clearer just how much of our genome we share even with the humble fruit fly, by the late 1990s Rose realized that it wouldn't be necessary to switch to mice. He could stay with his familiar flies and still work toward the goal of human life extension.

By comparison with Rose, many of his colleagues were even further away from the likes of Kurzweil and de Grey. As recently as 2002, fifty-one scientists working in the field produced a document warning that we shouldn't hope for the impossible. They claimed that there were no current therapies that could put off aging. The aim of this group was to help people avoid wasting time and effort on therapies that simply don't work, but in the process these scientists seem to have been unduly pessimistic.

One of the foremost critics of the "we could stay alive forever" approach to life extension is S. Jay Olshansky of the University of Illinois in Chicago. Physical immortality, says Olshansky, is seductive, but we should not fall for its siren song, particularly if we are attracted to the idea of living forever by the wisdom of the ancients and the mystical powers of the alchemists. He comments drily: "What do the ancient purveyors of physical immortality all have in common? They are all dead." Despite his rhetoric, Olshansky does accept that gerontologists "will eventually find a way to avoid, or more like delay, the unpleasantries of extended life," but he can't believe the extreme timescales that the likes of de Grey put forward or that Kurzweil's plan to live forever could possibly work.

Olshansky aims for a more practical-sounding seven-year increase in the average human life span. His argument is that at the moment the risk from age-related diseases rises exponentially throughout your life, with the chance of dying doubling every seven years—so by adding seven years to our average life span we would cut the probability of dying at any particular age in half. Seven years should be relatively easy to achieve if Richard Miller of the University of Michigan in Ann Arbor is correct. It has been known for some time that significantly reducing the calories in an animal's diet will increase its life span. Michael Rose had observed this in fruit flies, it is a well-known effect in mice, where the life-span increase is about 40 percent, and it has even been observed in human beings, for example, on the Japanese island of Okinawa, where conditions naturally imposed a very restricted diet.

We can't reasonably expect most people to live on a drastically low-calorie diet to extend their life, but Miller points out that studies on mice have now pinpointed ten mutations that result in the same effect as the calorie restrictions. This makes it possible to conceive of drugs that could mimic these mutations and cause extended life without the need to starve ourselves. If a similar effect could be achieved in humans as has been in mice, Miller believes that we could look forward to a healthy life averaging over 110 years.

Other scientists, notably Lenny Guarente and Cynthia Kenyon, who set up a commercial venture working on antiaging, focus on the related impact of insulin-like growth factor. This is a messenger chemical that is effectively a troublemaker if it gets in the wrong place. There is evidence that reducing its ability to link to certain receptors in cells extends life spans of a wide range of creatures. There is every reason to believe there would be a similar effect in humans if the right drug could be developed.

A third option (which lies behind some of the supplements in Kurzweil's first bridge) is to counter free radicals, the highly reactive chemicals that can easily cause damage to DNA. Although the body has sophisticated mechanisms to fix damage like this, as we have seen (page 43), some errors (mutations) do slip through, particularly in the limited DNA that resides within the mitochondria, the power units within the cell, which have genes that are unusually susceptible to mutation. This is because the chemical processes involved in managing energy for the cell produce a lot of free radicals as by-products.

It does seem that aging is influenced by free radicals, and cutting these down, the job of a group of chemicals called antioxidants, does appear to extend life. That's the reason Kurzweil and others, following the often-derided ideas of Nobel Prize–winning chemist Linus Pauling (also winner of the Nobel Peace Prize), recommend taking large doses of vitamin C, which is an antioxidant. Similar positive behavior is attributed to green tea and red wine. But although food that contains antioxidants may have a positive effect on reducing the risk of heart disease, hence increasing life span, it doesn't have the direct effect on life span that increases in the body's natural supplies of antioxidants seem to have. Both fruit flies and mice have been shown to live longer if their natural production of antioxidants is increased genetically, and drugs that operate this way are also under development.

Despite Olshansky's very reasonable doubts about the aims of immortality, we shouldn't entirely dismiss Ray Kurzweil's vision. While there is no certainty about *all* of Kurzweil's bridge one actions that lead him to his 250 supplements a day and intravenous treatments, we do have a pretty good understanding of what actions can be taken to put off (for instance) heart disease, indubitably increasing our chances of an extended life span. Some of the genetic work in bridge two is already possible and much is on the cards, including Miller's idea of mimicking the effect of the mutations that cause extended life in mice, while there can be little doubt that at least some of Kurzweil's nanobot dreams from bridge three will eventually come into place.

Michael Rose gives us a sober assessment of the realities. Aging, he says, won't be cured by the sort of single elixir of life that was beloved by the alchemists. It isn't a single process but a complex mix of problems in various systems within the body that are no longer particularly well regulated as we get past the age of reproduction. Once we have reached that sort of age, there's a reduction in the evolutionary pressures to keep our bodies functioning well. From the evolutionary viewpoint, once we have reproduced we really aren't necessary anymore. The biological processes that lead us to maturity and reproduction are controlled to make things happen in a certain way. Once the goals are achieved, the control is let go and we drift into old age and body failure. It's not a guided process but the gradual breakdown of many systems, so it's not just a simple matter of finding the right switch and throwing it.

This may dispose of the alchemists and their one-shot elixir but not that alchemist of the scientific age Ray Kurzweil, who certainly doesn't assume that we need a single silver bullet to solve the aging problem. Michael Rose looks forward hopefully to the future. He comments: "The prospects for discovering how to postpone human aging were negligible for all of human history up to 1980. From 1980 to 2000, those prospects were hopeful, but not outstanding. Since 2000, the year of the sequencing of the human genome, the prospects for postponing human aging have become excellent. We now have all the basic tools we need, except an organization with the willingness and resources to do the job."

Human beings have dreamed of extending their life span as long as we have been conscious of the fact that each of us as an individual is going to die. One huge breakthrough that hasn't really been covered here is the way that medicine has enabled us to move away from high levels of infant mortality. When we look back at average life expectancies in the Middle Ages and see an expectation of twenty-eight, as we have seen, this reflects just how many children never reached maturity. In the average sense, we have already done much to overcome this "natural" tendency. Our self-updating through better medical treatment and understanding of what conditions make for infant mortality has pseudoevolved us into a creature that is much more likely to survive to adulthood and so to live out a reasonable length of life.

Modern medicine and hygiene have taken us even further. Where typical adults of the Middle Ages who survived childhood might have made it to their fifties, now most will have another twenty years. Once we get over the obsession with an elixir of eternal life that drove alchemy for so long and really understand the way the aging processes work we can go even further. It seems there is every reason to believe, despite the naysaying group of scientists in 2002, that within the lifetime of many who are alive today, if perhaps not Ray Kurzweil, we should be able to extend human life significantly further. Death may not have lost its sting, but it will certainly be postponed. Our ability to delay mortality makes us into a different creature indeed.

Strangely, one of the biotechnologies most likely to influence extension of life, and one that is a classic example of our adapting nature to transform our own species, is one that seems to be about making copies rather than staying alive longer. That's cloning, the process of producing another creature (in the case we're interested in, a human creature) with identical DNA to the original. (The word "clone" comes from the Greek word for a twig, suggesting a branching off, producing multiple versions from a single original.) What has to be got out of the way immediately is the hackneyed *Boys from Brazil* image. This 1970s movie based on an Ira Levin novel involved multiple clones of Adolf Hitler, raised in the hope of generating a replacement of the Nazi leader and bringing back the Third Reich.

Leaving aside the fact that, even with today's technology, human cloning is impossibly difficult, so back in the seventies it was a nonstarter, there is a fundamental flaw in this concept, because a clone is not the same person as the original donor. Admittedly, the plot of this movie was better than many story lines that involve cloning, in that the clones did have to grow up normally rather than magically turning into adults overnight. And Levin tried to allow for the way the environment has an impact on the way a human being develops by insisting that the clones were brought up in a similar household to Hitler's, down to their fathers dying at the right age (not too great a prospect for the surrogate fathers, who were murdered), but this simply isn't enough. We know this because we already have examples of human clones brought up in the same environment and they still differ significantly from one another.

This sounds contradictory, having said that human cloning isn't currently possible, but these are natural human clones—identical twins. Even though they share the same DNA and are usually brought up in the same environment, identical twins are clearly different people by the time they are adults, often even looking significantly different from each other. (Technically identical twins are "more identical" than a clone is with its donor, as both twins have the same mitochondrial DNA—the small number of genes passed only from the mother—while a clone will have the mitochondrial DNA of the host egg.) The more common fraternal twins come from two separate eggs, so are effectively just siblings born at the same time, not clones.

Not only do identical twins have subtle differences in their environment that mean they will develop differently—they often have different friends, for instance—but they will be biologically different, too. Our genetic code isn't 100 percent fixed at the embryo stage. Each of us will gradually accumulate small changes, for example due to errors in copying DNA as cells divide. Different genes will be switched on and off because of subtly different environmental cues. The result is two distinct people. This "clone difference" has since been observed directly in cloned animals. The first cloned cat, for example, named CC, produced by Texas A&M University, was anything but a carbon copy of its parent. It had a gray-striped coat, where the donor cat had a coat with brown splotches on a white background.

These differences mean that even if you could produce a clone of yourself and accept the tortured argument that this somehow extends your life because this other "you" will still be alive after your death, the fact is that you are only really producing a brother or sister. And that "even if you could" is a big "if." When the sheep Dolly was cloned back in 1996 it seemed only a matter of time before someone would do the same with human beings. There were ethical considerations, certainly, but many assumed that what could be done with a sheep could also be done with a human. Some publicity seekers even claimed to have produced human clones, though strangely, they have never been able to produce the very simple proof of the truth of their claims. But the lessons learned from Dolly were more complicated than this. It's worth briefly exploring what happened to Dolly to understand this.

Cloning is reproduction not by combining half of the genetic material from each of two individuals, as in normal sexual reproduction, but by duplicating a single set of genes. In cloning an animal like Dolly a piece of genetic material is taken from the original host. In the case of Dolly's parent, the source was from the mammary gland, which is why the lamb was named after the singer Dolly Parton. As it happens, Dolly's "mother/sister" was long dead—the cell used was from a culture, kept alive in the laboratory, rather than taken directly from a living animal. The contents of one of the donor cells are used to replace the insides of an unfertilized egg cell, which has its normal contents sucked out.

A tiny burst of electricity, like the shock that brought to life the Frankenstien monster, was used both to help the nucleus fuse into the egg that would become Dolly and to give the process a kick start. The egg, implanted in a host mother, began to grow in the normal fashion, and after the appropriate period of time Dolly was born. She appeared to be a perfectly normal, healthy lamb. This, though, is the easy reading, "no snags" version of history. If cloning were that trivial, we would have clones popping up all over the place and the few individuals who say they have produced human clones would be proudly displaying them, rather than making grandiose claims but never producing any evidence. In practice, getting that far has proved hugely difficult.

Even producing Dolly took many years. Although it had been possible to use this technique with frogs for some time, it just wouldn't work with mammals. The breakthrough was to start with a cell in a different state from those that had originally been tried. Most of the time our cells aren't rapidly duplicating themselves as they do when a fetus is growing. All the original experiments had used cells that were in the right state to split. What the team at the Roslin Institute in Scotland, who came up with Dolly, tried instead was using quiescent cells, cells that had initially been splitting but then had their nutrients removed, so the growth process stopped. These proved effective.

The snags weren't out of the way yet, though. While the quiescent nuclei did seem to work when transplanted into eggs, most were false starts. Out of 276 initial tries, only 29 showed any sign of activation, and of those 29 implanted in surrogates, only 1— Dolly—lived. But surely, now we've had Dolly, it's easy to get better and better at the cloning business? Isn't it only a matter of time before we see those human clones?

No. First of all, although Dolly seemed perfectly normal, she died unusually young for a sheep. This has been attributed to old age, despite Dolly only reaching around half a normal sheep's life expectancy. (To be fair, most sheep don't even get that far—we eat them first.) Quite where this suggestion that Dolly died of premature aging came from isn't clear, as Ian Wilmut, the scientist behind Dolly, makes it clear that she died of pulmonary adenomatosis, an infection that isn't uncommon in adult sheep. She did have slightly unusual arthritis, though this was put down to her being overweight, thanks to too many visitors slipping her tidbits. According to Wilmut, "the post-mortem had revealed nothing particularly unusual for an animal of her age and weight."

However, it's certainly possible that a clone could die unusually young. One reason for this is that the clone's cellular clocks could be telling the clone that it is the same age as the donor. Chromosomes, the packages of genes that make up our genetic instructions, as we've seen, have little tags at the end called telomeres. Each time a cell divides, its chromosomes lose a bit of their telomeres. Dolly's telomeres started identical with those of her six-year-old parent. It seems possible that the older the parent animal, the less time the clone will have before the problems of old age set in, though Dolly's experience doesn't provide any evidence for or against this theory and many clones since Dolly haven't had these shorted telomeres, so in this respect she may have been an oddity.

Equally, clones' genes can get damaged in the rough-and-ready process of bringing a clone into existence. Cloning is a bit like trying to repair a delicate watch with a hammer and chisel— you can get lucky and fix things, but it's much easier to do damage. Later studies of animal cloning have shown that the process tends to modify the DNA, damaging important genes and resulting in the inability of many embryos to survive. Those that do live often suffer from serious problems. All the evidence is that these potential problems get worse with monkeys, worse still with apes, and it is quite possible that it may never be practical to safely produce a cloned human being without producing many damaged children. For this many people will breathe a sigh of relief—not for the potential to produce damaged children but because it prevents experimenters from undertaking the cloning.

Some have argued, in fact, that all cloning will always be an unsafe process that throws up many failures. Certainly the success rate so far is dismally poor with mammals—typical between 0.5 and 1.5 percent achieve a successful birth of an apparently undamaged baby animal—and although cloners are getting better at the mechanics of producing a cloned egg, they don't seem to be making much impact on this failure rate. Bearing in mind a good number of these failures take place after the embryo has started to form or even after the animal is born, this is a totally unacceptable risk if the same technology were applied to humans. No one with any moral standing would consider it acceptable to produce a clone if the known price was the production of tens or hundreds of deformed fetuses and babies as by-products.

The problems with safe cloning don't mean that you can't clone human cells, though, nor do they mean that cloning isn't relevant to the extension of life—in fact, the technique has a lot to offer, on two separate levels, in the form known as therapeutic cloning: using cloning techniques to overcome failures in the body. The process involves that other contentious biotechnology subject—stem cells. Rarely has a scientific topic had so much exposure in the speeches of politicians who haven't a clue about the subject than has been the case with stem cells.

Stem cell research, with all its accompanying debate and scandals, is one of these new lines of medical thinking, based on a better understanding of how the body builds itself and mends itself. All cells are not created equal—some are much more flexible than others. If you look at the cells in the body, they are not identical. This is pretty obvious. The cells in your skin are quite different from the cells in your hair, your blood, or your flesh. They look different; they feel different. Yet all your cells came from a single, original cell that divided over and over again as you were developed in the womb.

The very first embryonic cells to form are totally flexible. They can become anything. But as cells divide they begin to differentiate. As they become more and more different they become more specialized. It's a bit like children in education, starting as generalists but becoming more and more specialized as they get further down the line. After a number of divisions, a particular type of cell is only likely to split to make more of the same kind of cells. Cells that can become different types of cell are called stem cells.

So far so good. But there are two broad types of stem cells—embryonic stem cells and adult stem cells. (The word "adult" is a bit misleading; these are the stem cells that children as well as grown-ups have.) Adult stem cells are more specialized than embryonic cells. In principle an adult stem cell from a kidney could produce all the cells required for a new kidney—but it couldn't produce a new liver. Embryonic stem cells can do anything, can become anything in the whole range of a human cell's capabilities.

Without doubt, stem cells might be a real miracle tool for medicine. In the long term they might enable us to grow replacement organs and to treat cancer or repair damage to internal organs and systems. More short-term but still important possibilities are treatments for conditions that potentially shorten life span such as Parkinson's disease and diabetes. However, so far the only way to get those particularly effective embryonic stem cells results in the destruction of a human embryo. The extraction is performed at the stage when the embryo is still a collection of a few cells—it certainly has no nervous system, no brain, no ability to feel or be conscious or to suffer; nonetheless, such an action presents a real moral problem to many people.

Even so, scientists are often bewildered by the apparently illogical resistance to all stem cell research. For example, in 2006 President George Bush vetoed a bill to enable limited embryonic stem cell research. This would have made use of cells from excess embryos created during IVF treatment for infertility. These embryos are normally destroyed—it's hard to see how destruction is a better outcome than making use of the cells to develop therapies that have the potential of being lifesaving on a tremendous scale.

Cloning comes into life extension because the body is very good at destroying invaders. Our immune system is designed to spot foreign material, like bacteria, and to destroy the incomers before they can do too much harm. But when using foreign cells or organs for therapeutic reasons, we need the body to accept them. The immune system has to be fought into submission, which puts the patient at risk of infection, and there is always the possibility that the immune system will fight back and reject the implant.

However, if it were possible to start with cells that were clones of the patient's own cells, then there would be no rejection—the new cells would be recognized as "one of ours." So there is a huge amount of interest in therapeutic cloning—the production of cloned stem cells that can be used to help repair the original donor. This therapeutic cloning is banned in many countries and remains controversial, not because of its application but because the source of the cloned cells is, in effect, a very early embryo that in principle (though not in practice) could be allowed to develop into a living human being.

Acceptance of the procedure was not helped by one of the biggest scandals ever to rock the scientific world. A South Korean scientist, Hwang Woo Suk, who claimed to have made big steps forward in the cloning of human stem cells, was shown in early 2006 to have faked all his research. Hwang was disgraced, and major journals, such as *Science,* that published his results were highly embarrassed. The whole process was thrown into temporary disrepute. Stem cell research and therapeutic cloning is not going to go away, but it did receive a significant setback.

There is hope, for those who have doubts about the use of embryonic stem cells, of using adult cells instead. A technique called transdifferentiation, or human somatic cell engineering, has shown promising capabilities of taking one type of cell and converting it into another, without the use of embryonic stem cells. With this technique, easily replaced cells like skin cells could be converted into cells required to repair or replace an organ—and as the process uses your own DNA, there would be no question of rejection. But this technique is not proven and may be of limited use.

Cloning, then, is another weapon in the armory of those who hope to extend human life by replacing the failing cells and organs that eventually bring life to an end, however healthy the individual's lifestyle, but there are those who say that prolonging life isn't desirable even if we have the technology to do it. Some religious figures believe that to do so is to interfere with God's will. They argue that by extending our lives we will continue beyond our allotted spans. Given the subject of this book, this seems a spurious argument. It is fundamental human nature to upgrade ourselves—if you believe in a creator, then that creator built this into us and we are fulfilling that design by trying to extend our lives. Anyone who doesn't believe we should be trying to extend lives shouldn't make use of any technology whatsoever: it's all part of the same process.

Others worry that we would get bored with a longer life (see page 267 for more on this)—but they don't really have a realistic argument on any practical timescale that is envisaged by anyone other than Ray Kurzweil and friends. There are few people indeed who wouldn't enjoy having a little more time to do things they never got around to. Not many of us say, "My life really is a bit too long."

All the modifications we have looked at so far (with the exception of the more extreme aspects of Kurzweil's third bridge) have involved ways to keep a human body going longer but in much the same way. However, this isn't the limit of the possibilities for upgrading this way once we do have a better control of genetic mechanisms. Evolution is often a response to dealing with a changed environment, as seems to have been the case with the original development of *Homo sapiens.* As we get better at manipulating genes, we will have the ability to modify the human form itself to live better in different physical environments—perhaps even environments that are totally alien to us at the moment.

This has already been begun with plants. Researchers at Texas A&M University, for example, have identified genes in tomato and cotton plants that can be modified to improve the plants' ability to withstand drought. But as we've seen, making changes to individual genes without the nanobots that Kurzweil envisages is a haphazard business. It isn't a matter of getting into the DNA and changing individual letters of the code but rather taking a whole chunk of DNA and getting it to splice somewhere into the existing sequence, with no control over exactly where. It's this inaccuracy that gave early attempts at gene therapy, like the 1999 Paris trial where three children died, a bad name. There just wasn't enough control.

It is only as recently as 2006 that genetic engineers such as Paul Eggleston of Keele University in the United Kingdom have developed tools to enable precise changes to be made to the DNA of living cells. These would have been a surprise twenty years earlier. A mechanism was found back then that enables precision manipulation of mouse genes. By combining an embryonic stem cell with a new piece of code and a piece of existing mouse DNA, researchers found that occasionally the mouse's internal systems would take this change as a template for action and would replace the matching existing code with the new piece. This means that mouse genes can be manipulated with ease—but to everyone's surprise, the technique wouldn't work on other animal cells. It proved to be a one-off solution for mice.

Recent breakthroughs have made this sort of control equally possible in other creatures, including, potentially, humans. One approach uses special chemicals called recombinases, the molecules that viruses use to help them target a position for their attack. This approach is hit-and-miss to begin with, as it's necessary to get a special "target" piece of DNA in the place you want to add in new code, a target that still has to be gotten in place using the old random methods—but given that placement, you can then deliver code reliably in the future, and Eggleston has already used this technique in work on a genetically engineered mosquito that can't transmit the parasite that causes malaria.

A more sophisticated approach still, more suited to human gene therapy, is under development using the elegant miniature tool of molecular scissors. These are a combination of a special molecule designed to lock onto just one part of the DNA structure and enzymes that can cut through the DNA chain. (So really the "scissors" are more like a molecular chain saw with Velcro attached.) If you cut DNA at the same time as introducing the new code you want to add, the cell's own structures that try to repair the DNA tend to incorporate the new code at the cut point, making for precise insertion.

Again, this isn't a perfect solution. The molecular scissors don't always hit the right spot, and there is some concern that an improperly repaired break in a DNA strand could lead to cancer outbreaks—this hasn't been found to happen in any known cases, but equally it hasn't been shown that this won't occur. There is more research to be done. Yet it's clear that we are likely to be able to manipulate human genes, both for therapy and to cope with different environments, in a relatively short time.

With full ability to modify genes it isn't too far-fetched to imagine being able to tailor human beings to cope better with the heat of global warming or low-oxygen atmospheres (though as yet the idea of producing "fish men" who can live underwater is still the realm of science fiction). According to the Palo Alto–based Institute for the Future, a respected agency used by the British government for horizon scanning, "[w]e will dramatically alter, enhance and extend the mental and physical characteristics that nature has dealt us . . . we will remake our minds and bodies in dramatically different ways." Most scientists working in the field feel that this is an exaggerated viewpoint, but tailoring to be able to cope better with heat or pollutants in the atmosphere seems a more likely possibility than transforming humans so drastically.

The urge to preserve and extend our lives is an obvious driver for the unnatural selection provided by human upgrades. Yet this is not the only evidence we have of the way our ability to shape ourselves has gone beyond nature. Just as natural selection has resulted in sometimes bizarre modifications to attract the opposite sex—a peacock's tail, which puts the peacock at huge risk from predators in exchange for appearing sexy to peahens, is an obvious example—so the same pressures led early humans to improve on their natural allure.

3.
Cosmetic Charisma

"Zaphod, great to see you, you're looking well. The extra arm suits you."
—Douglas Adams, *The Hitchhiker's Guide to the Galaxy*

To keep the species alive it is essential to be attractive to the opposite sex—and we are prepared to go to extraordinary lengths to do so. Even some animals indulge in hard work to make themselves more desirable. The peacock may not work for its tail feathers, but the dull-looking male bower bird makes up for its unexciting appearance by building a treasury of exciting objects with the sole purpose of attracting the attention of the female of the species. Admittedly this decoration isn't worn, but the effect is much the same as a cosmetic makeover. Human beings, though, leave even the bower bird well behind.

As our ability to upgrade our appearance has developed, so have the effort and expenditure that have gone into cosmetic modification. You only have to watch the relentless TV advertising to be aware of this. It has been estimated that the noninvasive face-lift and facial rejuvenation market (that's facial modification without surgery) *alone* will amount to over $4 billion in 2010, while the total worldwide cosmetics market was worth around $116 billion back in 2005. This is serious business, and that's without even considering the biggest cosmetic market of them all, clothing. As we've already seen, clothes may be great to keep us warm and protected, but that's not the only or perhaps even the main reason we wear them.

It is difficult to establish now just how early humans began to adapt themselves by giving themselves an artificial outer skin to look more attractive, to display status, and to keep warm. The materials likely to have been used in early clothing—animal skins and vegetation—would have long decayed. It's a similar problem to the one scientists face when trying to re-create the world of the dinosaur. Anyone who has seen *Jurassic Park* or the TV show *Walking with Dinosaurs* or who has "met" animatronic dinosaurs at the local museum may think that they know what dinosaurs look like. In fact, all they have seen is a guess—well educated, it's true, but a guess nonetheless.

The remnants we have of dinosaurs are the parts that don't rot away—primarily bones that have fossilized into stone. The skin is long gone; for all we know, dinosaurs could have been purple like Barney (though most paleontologists would be careful to add that this is highly unlikely). Reconstructions of the appearance of a living dinosaur are based on what we know about existing animals that are in some ways similar and so may have similar skin—lizards, for instance—but there is a considerable presumption in taking that step.

Similarly, we know that at some point the early humans managed to adapt their natural weak skin and boring appearance with clothing, but anything they might have used has long since rotted away. Entirely disappeared. Even so, these transient materials have managed to leave behind some clues to show that they were once here.

Woven cloth, we know, dates back at least twenty-seven thousand years, because clay of this age has been found at the ancient settlement at Pavlov in the Czech Republic, with the imprint of cloth on its surface. Just as dinosaur footprints have been found preserved in clay that subsequently turned to stone, here the medium holds a message of another passing item. Stitching, to fix together smaller pieces of cloth or animal skins, we can trace significantly further back. The oldest bone needles that have been found, at Kostenki, a village in Russia, date back around forty thousand years. But perhaps the best clue to when the first humans began to upgrade their skin comes from the humble louse.

Lice are evil-looking little parasites. They are specialist bloodsuckers that live on their host's skin, taking sips from the blood beneath. The most familiar human louse in the modern world, the head louse (*Pediculus humanus capitis*), is very fussy about sticking to its preferred environment around the head hairs. Yet eventually a different form of human louse was to evolve from the head louse. This is *Pediculus humanus humanus*, the body louse. From the genetic makeup of this parasite, scientists at the Max Planck Institute for Evolutionary Anthropology in Leipzig, Germany, have estimated that the body louse came into being between fifty and one hundred thousand years ago. This timing can be worked out by looking at the variation between DNA sequences in the two creatures—the more difference, the longer ago the division between head and body lice occurred.

Why the interest in these unpleasant little pests? It is argued that the reason the modified body louse survived and thrived away from the protection of head hair was the introduction of clothing. When we began to wear clothes, there was a new environment lice could make use of—before then, the uncovered skin had been too exposed. Interestingly, this timescale corresponds well with the timing of the move of humans out of Africa into colder climates, which could have been the spur that brought on the use of clothing.

Older garments are likely to have been based on animal skin, whether simple skins or hides stitched together with sinews, but the more flexible manufactured material came in surprisingly early. As we have seen, there appears to have been some woven material as early as 25,000 B.C., and another clear indication of a manufactured garment is found on a carved bone statue from Lespugue, France, depicting a female figure that wears what has been described as a skirt, apparently made from twisted strings. The term "skirt" is perhaps misleading—the garment hangs from a long string attached to the waist, and is suspended at the back of the wearer, below her bottom. It has been reasonably suggested that this "skirt" has more symbolic or eye-catching purpose than having any function as a practical garment. But it is a manufactured piece of clothing of sorts, and the statue dates back to 20,000 B.C.

Amazingly, given the fragility of plant materials over time, there is a remnant example of textile, again from France, that is seventeen thousand years old. Found in the famous Lascaux cave with its vivid wall paintings of hunting scenes, the remnant is of a piece of string formed by plaiting together three thinner strings of twisted fibers. It was found by accident, embedded in a lump of clay, and though the original plant material was decayed too far into a messy remnant of carbon ooze to be able to work out what the string was made from, the clear imprint in the clay made it possible to deduce just what the original cord was like.

It seems, then, that the earliest enhancements of our naked bodies were animal skins—a natural approach to give warmth that must go back as far as hunting—and then knotted fiber-based material and weaving. Until very recently, wearing fur had huge prestige (and even now, despite fervent animal rights protests, it is showing signs of coming back). In part this reflects the exclusivity of expensive furs such as mink and sable, but it is also an echo of the dual benefit that must have been seen in fur in the early days.

Ancient fur coats don't survive, but needles to sew hides together and scrapers to remove the remnants of flesh from the skins do. Without doubt furs brought warmth and physical protection. The skin of most animals, especially after treatment, is stronger than our own. Even without the fur on, this is useful, as when leather is formed into shoes. With the cheapness of artificial materials, leather shoes wouldn't be popular today unless they were more pleasant or desirable to wear than the plastic equivalents. And the warmth of the fur in an animal skin itself adds benefit. Fur is great for trapping air to act as an insulating layer.

Yet it is also likely that the early wearers of furs saw them as giving something more than warmth and protection. If you wore a fur, then you (or your mate or the parent who provided the skin) were a successful hunter. It also was likely that the fur of a powerful animal would be seen as giving some of that power to the wearer. Sympathetic magic—gaining some characteristic or link because of appearing like something—is a very natural assumption, even though it makes no sense scientifically. "If I wear the coat of a bear or a wolf, then surely I become more like a bear or a wolf" has no logical basis as an argument, but it carries plenty of emotional weight to add to the practical benefit of the garment.

Nonwoven fiber-based clothing, like the string skirt, does not require as much preparation as getting threads ready for weaving. Before a woven cloth could be produced, natural fibers had to be extracted from long plants such as vines and other vegetable matter, notably hemp and flax. The plants would be dried, then soaked in water so that the fleshy parts started to rot, a process known as retting. The disintegrating plant would then be twisted and pounded to separate off the woody parts, before the fibers were combed out.

At this stage, the fibers would be limited in length. To weave anything of a practical size it is necessary to spin the fibers together into a longer thread. The fibers would first be joined together, often by simply wetting them with saliva (enzymes in the saliva partly break down the vegetable material to help it bond) before twisting the fibers together. The crude thread then had to be spun—given a twist to produce a stronger multilayered strand. Initially this was done by hand, using a simple spindle—a stick, usually passing through a wooden or clay weight, which helped keep the spindle rotating evenly, rather like a flywheel.

The spindle would be set spinning by rolling it down the thigh and was allowed to drop to the ground, leaving a length of spun thread between the hand and the spindle that could then be wound up before the next length was spun. Usually thread for weaving would be multi-ply, with two or three threads spun together for strength.

Crude weaving, interlacing the threads to turn them into cloth, can be done by hand, but to make anything larger than a small patch of material requires a loom—a framework to support the threads as they are woven together. It is known that at least two forms of looms were in use in the late Stone Age and early Bronze Age—a horizontal ground-based loom, originating in the Middle East and North Africa, which used wooden sticks to support the weave, and a vertical loom that dangled from a beam with the warp (the straight threads that the other threads are woven around) hanging downward, pulled down by weights, so that the weft (or woof—the other threads) could be threaded through. This style of loom seems to have originated in Europe. Originally the weights were usually made of clay.

The thread itself would originally have been plant material like the early strings, particularly the flax used to make linen, but first in Europe and later throughout the world wool was increasingly used for its extra insulating properties. In the Far East, silk, with its extremely long, smooth threads, was soon employed.

The lengthiest part of the process, before mechanization like the spinning wheel, was the initial spinning, taking the short natural threads and twisting them together into a much longer and stronger spun thread. This could take up to ten times as long as it would take to incorporate the thread into a woven sheet of fabric—it was a slow but necessary manual process, particularly important when moving from plant fibers, which could naturally be quite long, to wool (and later cotton), where weavers have to contend with filaments that can be as short as an inch in length.

Wool seems to have come into use around six thousand years ago, following a degree of selective breeding to get sheep with the best coats for fibers suitable for spinning. The need for comfort and for practicality would also lead to a pattern of clothing layers that is still common today. Although wool is warmer and easier to dye than a plant-based fabric such as linen, which is based on flax, it is scratchy and uncomfortable on the skin. Because of this, inhabitants of European countries that took up the use of wool tended to wear a woolen garment over a linen tunic or shirt, just as we do now. The linen layer was more comfortable and could be washed more easily.

The first properly preserved garments we have are from the Bronze Age: these include a set of hunting clothes that are around fifty-three hundred years old, a five-thousand-year-old linen shirt from a First Dynasty tomb at Tarkhan in Egypt, made with surprisingly modern styling, and a female outfit from three and a half thousand years ago. This newest set of clothes mixed string-based and woven garments, showing that the use of string carried on after the development of woven cloth. In fact, the string skirt, found in the Danish Bronze Age settlement at Egtved, is a garment that wouldn't look out of place on a modern catwalk.

Located in South Jutland, Denmark, the skirt was on the remains of a sixteen- to eighteen-year-old woman known as "the Egtved Girl. The woman's body was preserved in a log coffin that had soaked in acidic groundwater, reducing the usual corruption of time. Her wraparound string skirt passed twice around her body. String after string dropped around fifteen inches from a waistband, ending in an attractive fringe of knots. This early miniskirt was accompanied by a short-sleeved woolen bodice and a woven belt, and the body was laid on a cowhide, covered in a woolen blanket.

The oldest of the Bronze Age clothing finds was the attire of the famous "iceman" from the border between Austria and Italy, known as Ötzi. Dating back fifty-three hundred years, Ötzi was an accidental discovery by tourists in 1991, who thought at first he was the victim of a recent accident. Frozen in a glacier, he was surprisingly well preserved and wore leather leggings, loincloth, and shoes, a fur cap, and a woven grass cloak.

Many of the earliest samples of cloth that have survived are decorated with patterns, beads, and color. Even thousands of years ago, there was no question that clothing had a decorative role. From Stone Age figures from Hungary, where similar patterns appear both as skin markings and in clothing, it seems likely that the patterns started as painted body markings (see page 91), which were then adapted to the new medium. This became easier with the addition of wool to the textile makers' armory—wool is easier to dye than the plant fibers that had dominated weaving before and comes in a natural range of colors without even any necessity to dye it.

So we have a picture of how clothing was developed, but we still haven't clearly established why humans wear clothes at all. As far back as 1929, the French-American artist and anthropologist Hilaire Hiler suggested that warmth and physical protection was rarely the prime motivator in adopting clothes. Some of Hiler's observations are difficult to repeat today, as cultures have changed—and it's hard at this distance to separate fact from supposition, when many early anthropological studies were corrupted by the ease with which the scientists were fooled into believing everything they were told.

However, Hiler may have had a point when he observed that "the tribes of central Australia wear no clothing although they often suffer from cold" and also that African safari porters would work in the heat of the day wearing every layer of clothing they could manage, then relax in the evening without their clothes. The implication was that wearing these clothes gave the porters an appearance of power by association with the Westerners they worked for that outweighed the discomfort of being overdressed for the weather. Wearing clothing—or not—had, according to Hiler, a more important significance than comfort.

It is certainly true that many people do still wear clothing that is designed more for status or for attraction purposes than it is for practical benefit. The bizarre white wigs of the British judicial system are no longer necessary to cover shaven heads but serve to mark a certain status in the court. Teenage girls who go out in winter displaying bare midriffs are not acting sensibly in terms of wearing clothing for protection—they have a different end in mind, both marking their status as young and available and catching the eye of the opposite sex.

Yet it would be folly to argue that status or sexual attraction is the sole reason that many of us wear clothing. The things I wear as a writer who works at home are usually chosen for practicality. I will put on what's necessary to give the degree of warmth I need for the prevailing temperature—sweaters in the winter, T-shirts in the summer—and also for the degree of protection I need from the environment. That means, for instance, jeans rather than shorts, both because when I walk the dog I need my legs protected and also because I spend a lot of time writing, seated on a leather chair, which would be downright uncomfortable without the long pants.

Even so, throughout history marking out status has been a crucial application of clothing, from the ceremonial robes of the priest to the string skirts worn by the Egtved Girl and the twenty-thousand-year-old French statue. These skirts were clearly not to keep their wearer warm—in the older example the position made it obvious that this was not a functional garment, while the Danish woman had woolen clothing, which would have been much more effective than the swirling string skirt. Instead it seems likely that these string garments may have marked their wearer's status, or crossed over between a status marker and clothing designed to attract the opposite sex, where wearing a garment that gives hints of what is beneath has long been seen as more attractive than being fully covered or fully revealed. Interestingly, string "aprons" were worn (over other clothing) well into the twentieth century in parts of Central Europe by girls who were engaged to be married.

There is a long history of using clothing that indicates the sexual maturity of the wearers or whether or not they are available for marriage, from the Islamic *hijab* and medieval Christian wimples to the squash blossoms worn in the hair by unmarried Hopi women and wedding rings. The two applications for clothing of attracting the other sex and clarifying status emerge from a single natural source. In nature, animals use display characteristics both to attract potential mates and as indicators of status. These can evolve to the extent of being dangerous to the creature's survival—the peacock's immense, brightly colored feathers do it no favors when it comes to surviving an attack—but such is the importance given to the ability to sport visually attractive signals and to show higher status than other similar creatures that the dangers are biologically accepted.

In the natural state, we human beings are limited in our ability to display for the opposite sex, and from early times clothing proved an effective way to distinguish individuals that would far outweigh simple variability in hair color and length or in eye color. Similarly, special clothing became a clear way to glorify royalty, to pick out the priesthood, and to identify those who were exceptional because of the job they did.

It can be no surprise, then, that as clothes became more sophisticated—and looking back, for instance, at the Minoan dresses from thirty-five hundred years ago, they could be just as complex and styled as garments on a modern catwalk—color and pattern played an increasingly important part in their design. Basic natural dyes were common those thousands of years ago (interestingly, blues and reds were the easiest of the early dyes to fix and make run-proof, which is why many countries' flags sport some combination of red, white, and blue). Metal ornamentation, to add extra eye-catching appeal, goes right back to the earliest working of metal, well before the Bronze Age, when the soft, eye-catching metals like gold that could be easily worked were added to costumes: a very direct attempt to rival the peacock.

The use of the precious metals wasn't limited to rulers and other important people, but inevitably they tended to have more wealth and could sport more of these "standout" decorations. (This is in contrast with twenty-first-century leaders, who are largely rather gray and uninteresting in their dress.) Special forms of metal decoration, like the crown, were developed to highlight an individual's importance, while some colors were given special significance. Purple, a very expensive dye before artificial colors were introduced, was limited by the ancient Greeks to very important people and was restricted by the Romans to the emperor alone. This Roman tendency has carried on into the Rome-centered Catholic Church, which also has very specific color restrictions on costume, with the pope's white and the cardinal's red.

Increasing our appeal to the opposite sex, probably the most powerful driver of the three reasons for wearing clothes, was from the beginning a combination of highlighting, enhancing, and concealing natural features that the individual already had, depending on the look that was regarded as most attractive at the time. Highlighting would pick out an aspect of the body and bring it most directly to the attention. Enhancement attempted to make a particular area look (even) better than it naturally did, while concealment covered up a perceived defect. The string "skirt" on the twenty-two-thousand-year-old statue from Lespugue, France (see page 78), is an example of the first of these. It clearly was not for warmth or concealment but rather drew the eye to the sexual region.

The codpiece—a cloth structure worn over the male genitals—went from one category to another. It began life as a concealment when in Tudor times the length of the male tunic was shortened, revealing tights that were open at the top, leaving absolutely nothing to the imagination. Over time, as the design of tights and breeches changed, the codpiece ceased to be a functional piece of clothing but instead set out to enhance or augment the visual size of the male anatomy with padding, often to a ludicrous scale. Generally in a much more subtle way, much of the makeup, including the modern concealer, sets out to hide a feature that is considered unattractive.

By nature, human beings have relatively few ways to visually attract the opposite sex. We are bland animals to look at. It used to be a common racial insult to say "they all look the same" about whichever ethnic group you were insulting. In fact, it's a truth across the whole of humanity. To a nonhuman, we all look pretty much the same. Compared to birds with their plumage, for instance, we have very few distinctive features to offer visually. Hair, eye color, skin tone, face shape, and body shape are the main variants that we have available to us without any assistance, but the human urge to adapt would soon give us opportunities to give off signals that would take evolution millions of years to catch up with. Plumage like the peacock's took many hundreds of generations to emerge from a bland, brown bird—but human self-decoration can transform the appearance in a matter of minutes.

There isn't a part of the body that we haven't modified over the years, whether by changing its shape, transforming its color, giving false information about its size, or even altering its texture. From breasts to skin, from eyes to the whole silhouette of the body, the fashion of what is attractive or gives status has given us scope to make changes. Much was possible by taking our natural attractive features and making them look more like the accepted ideal, but we are creative animals, and however enhanced, the basic human form isn't enough. From surprisingly early times, enhancements became more dramatic than any human could naturally aspire to, and an obvious canvas for the art of self-improvement was the skin.

Without the plumage of birds or the varieties of patterned fur of most mammals, our skin is, frankly, dull. The best we can naturally achieve is to modify the coloration of our skin by varying the amount of sun our body receives. The skin is our largest organ and weighs about six pounds. The skin is not just the outer layer that you can touch, the epidermis, but continues below in the thicker dermis layer where your hair roots sit.

The very outer layer of your skin is dead. It's from here that the tiny flakes that become most of the dust around your house are produced. Immediately below this stratum corneum (like the cornea of the eye, the word "corneum" comes from the Latin *cornu,* meaning "horn," indicating its relative toughness) are two layers of cells, the squamous cells, also called keratinocytes, and the basal cells. It's these basal cells that rise to the surface and die as the outer coating, but they also play host to a different kind of cell, melanocytes, that give us our skin coloration.

Melanocytes produce melanin, the pigment that gives our skin its hue from the palest pink to the darkest ebony. The more melanin, the darker the skin and the less ultraviolet is allowed through. Our default skin condition has evolved to match the amount of ultraviolet in the sunlight of the regions we live in. Ultraviolet light is energetic enough to do damage to DNA if it can penetrate the outer layers of the skin and smash into the nuclei of our cells. This damage can result in cancer (melanoma).

In northerly regions there is very little ultraviolet that gets through the atmosphere in the winter and only a relatively small amount, though growing quickly with climate change, in the summer. Humans with a long family history of such low ultraviolet exposure have tended to lose melanin from the original African levels of our common ancestors, as some ultraviolet has to get through in order for the body to make vitamin D.

This leads to paler skin tones in these northern areas, with the extreme of the range being the delicate Celtic skin, which has very low levels of melanin. This is less dramatic than albinism, a genetic condition leading to a total lack of melanin and very white skin. (Melanin is also responsible for the brown eye coloration, so albinos often have pink eyes.) The melanin in pale skin sometimes clumps together to make either short-term darker patches (freckles) or more permanent clusters (moles).

Even in northern climes, though, the amount of ultraviolet

hitting the skin varies, and we can't rely on the timescale of evolution to deal with these month-to-month fluctuations, so the skin has a mechanism—tanning—to deal with increasing UV levels. When the skin is exposed to sunlight the melanocytes speed up production of melanin, so the skin darkens, absorbing more of the ultraviolet before it can do any damage to the lower layers.

We all have the same numbers of melanocytes, but they are naturally more active in those with darker skin: there is no difference biologically between a tan and dark skin. The mechanism, as is often the case with the body, is surprisingly subtle. The melanocytes produce small clumps of melanin, which are packaged up in little structures called melanosomes. These are then passed around to other cells, where they sit around the nucleus, protecting the DNA from damage when the ultraviolet comes crashing in.

Unlike some visible attributes, notably symmetry in the face, which seems to be universally considered attractive, there is no natural inbuilt preference to what is considered attractive in skin color. For nearly two thousand years in Western culture it was considered attractive to have as pale a skin as possible, because tans were a mark of those who worked in the fields. Women went to great lengths to protect themselves from the sun and used light-colored makeup to increase the effect of being "interestingly pale." In the twentieth century this was turned on its head. A heavy tan was considered healthy and attractive. Tans on those with pale skin are still often considered an appealing feature, though concerns about skin cancer seem to be reducing the popularity of tanning.

The bland appearance of the skin may mean that it is limited in its natural display to a decision on how much tanning is fashionable, but the relatively uniform coloration and large area of the skin make it ideal as a blank canvas for art, and it seems likely that body painting and tattooing were practiced from early times. Like the old fabrics discussed earlier, human skin does not usually survive for too long after death, except where mummification has desiccated it, so although it is assumed that some form of tattooing as a permanent version of skin painting was practiced thousands of years ago, the oldest known example of a tattoo belongs to the so-called Ötzi, the iceman (see page 83).

Ötzi has a total of fifty-seven tattoos, mostly consisting of simple lines. For remains of pictorial decorative tattoos we have to come forward to around 400 B.C. A body found in the Altai mountains of central Asia, locked into the permafrost, was decorated almost completely with images that may be monsters (though we have to bear in mind how different, for example, an Oriental lion image is from a Western version, so what we see as monstrous may not have been intended to be so) and images of fish and rams.

Body painting is likely to have been practiced even earlier than tattooing, because it requires a lower level of technology. Although evidence is hard to come by because of the ephemeral nature of the painting itself, there is no reason to doubt that body painting goes back at least as far as rock and cave painting, which can be traced to around 40,000 B.C. Paints used in cave art tended to be based on easily found pigments like ocher, an orange, brown, or red iron-based mineral, and it seems likely that early body paints were also clay and mineral based. Similarly, charcoal often features in cave art—again, such easily obtained natural pigments were likely to have been used for the first patterns on the body. It's hard to use charcoal without getting marks on the hands: it would have taken little stretch of the imagination to extend this unintentional marking to body decoration.

One of the best-known historical examples of body painting was the use of woad by the ancient British. Woad is a blue dye that was extracted from the plant woad or, more properly, the *Isatis tinctoria*. Sadly, this particular usage may be mythical. Woad was certainly used to dye cloth—in fact, it was the only blue cloth dye in Europe until Elizabethan times—but it wouldn't work well as a skin coloring. Although it is likely that the ancient Picts and Britons did use some form of body paint ("pict" comes from the Latin, meaning "a painted people," and Julius Caesar notes the habit of the Britanni of marking their body with some kind of paint), the chances are that the paint they used wasn't woad. Exactly when the practice died out isn't certain, though it seems not to have been still around by the time of the Norman Conquest in 1066.

Body painting continues in many cultures—perhaps most widely in the use of henna decorations on the hands and to provide temporary equivalents of tattoos—but it has never been quite as culturally universal as adorning the face, the part of the body that gives the best insight into our thoughts and hence our inner being. Painting the face has been the key to cosmetic enhancement from the very earliest of days. Whether the earliest makeup was an attempt at camouflage, like the face painting still used by soldiers, or had some connection with rites of passage or spiritual protection is not clear—but it does seem very likely that it didn't take long to realize that our faces could be enhanced for the attraction of the opposite sex.

The face is not just our main sensory region for exploring the world. Concentrating sight, hearing, smell, and taste in one compact area, it is the most direct route that others can take to explore our personalities. Apart from cultures where individuality is subjugated, the face is the only part of the body that is pretty well permanently left uncovered. A very high percentage of our communication comes from the face. However self-controlled we are, a lot of what we feel "leaks" out. This isn't always at the conscious level—it doesn't have to be anything as obvious as a smile or a grimace, though these are hugely important, and unlike language or hand signals, these facial expressions are universal indicators across humanity. The people watching our face may not even realize they have seen something in an expression but will still unconsciously read a feeling and respond to it.

This makes the face both very important as a means of attracting the opposite sex and also something that we like to change in appearance, either to conceal our feelings or to exaggerate them. Facial makeup is a critical tool in appearing somewhat different from what we actually are. Actors wear makeup because the powerful lights of stage and studio wash out their complexion and make it hard to read their expressions—they want to appear as if they were in a more natural environment. Older women, and to some extent men, have for some time used makeup in an attempt to conceal the ravages of time.

This attempt to conceal aging isn't a purely modern phenomenon, though it was much less usual outside the elite until very recently. In the eighteenth century, when pale skin was still fashionable, elderly ladies piled on white paint in an attempt to hide their wrinkles. But this doesn't mean that painting the face has always been socially acceptable—quite the reverse. The image of a "painted woman" is not intended to be a compliment, and in

Victorian times it was not unusual for women to pinch their cheeks and bite their lips to substitute a painful natural coloration for the forbidden fruit of the painted beauty they desired.

Any attempts to explore the earliest forms of face painting are restricted by the ephemeral nature of the medium. As we have seen, skin does not usually last long after death, and though mummified skin will hold the permanent marks of tattooing (see page 91), most painting disappears (and any paints used to decorate the corpse are not necessarily typical of makeup worn in life). Instead we are reliant on paintings of painting—the illustrations of makeup in art that has survived—and written descriptions once writing came into play.

The earliest civilization for which we have a lot of detail of their use of makeup is ancient Egypt. Everyone has seen the Egyptian wall paintings showing made-up faces with a thick black eye lining that continues back toward the ear as a straight horizontal line, with strong, clearly emphasized eyebrows and with what appear to be reddened lips. While there is some indication that this facial makeup was overemphasized in Egyptian painting, which was itself very stylized, it is certain that it was the norm for both men and women above the slave class to wear eye makeup, and a highborn woman would put every bit as much trouble into getting her facial cosmetics right as a Hollywood star (and would have had even more attendants).

Perhaps also only like some actors now, the Egyptian beauty queen would have a shaved head as part of her cosmetic treatment, wearing a complex wig to provide the perfect setting for that made-up face. Such was the importance of makeup at the time that for royalty it was an essential part of their royal insignia. As well as a crown, royalty, even queens, wore a short beard—like the wig, a fake that was fitted for special events by attendants.

Eye makeup does seem to have had particular importance to the ancient Egyptians, and the oldest artifacts in existence related to cosmetics are Egyptian palettes for eye makeup that are around twelve thousand years old. But eyes were certainly not only the target of Egyptian makeup artists. Skin would be made softer, more supple, and more pleasant smelling with perfumed oils, while makeup was also applied to cheeks, lips, eyebrows, eyelids, nails, palms, and the soles of the feet, as well as being used to accentuate the eyes.

The classic eye makeup, both the lining around the eye and the extra emphasis given the eyebrow, was most often in stark black, using a material called kohl. In the most common form of this traditional makeup component, the blackness came from soot, which was often mixed with minerals like galena (lead sulfide), stibnite (the ore of the element antimony), and pyrolusite (manganese dioxide). Kohls with brighter colors would use different chemical constituents—copper based for blues and greens, for example. At its simplest, kohl was applied by moistening a finger, picking up some of the powder and painting it in, though the Egyptians had special kohl sticks for applying it, made from a wide range of materials.

It is possible that the origins of using kohl around the eyes were more for practical purposes than for attraction, as an attempt at a solution to glare from the bright sunlight. Other cultures that have used kohl eye decoration—in India, for example—consider a dark lining around the eye to be effective in cooling the eye and protecting it from the sun. There is an element of truth to this—the bold black eyeliner of Egyptian times would slightly reduce the sunlight reflected toward the eye, though the effect it was countering would be very small compared with direct light.

Unfortunately, for a treatment intended to be healthy, the use of kohl presented a health risk. The use of heavy metal minerals like lead sulfide day after day could result in a poisoning through the skin, as the heavy metal content was gradually absorbed into the bloodstream. Kohl was also used to darken eyebrows—to give more control, it was not uncommon to shave off the eyebrows and replace them with fakes painted in kohl (much as eyebrows are now sometimes replaced by tattoos). Once again, unfortunately, this move was counterproductive because of the hot climate, where the eyebrow has a very practical role in keeping sweat out of the eyes.

Eye shadow was nowhere near as common as eyeliner but certainly was deployed in ancient Egypt. When it was used, it was also made from kohl, but one of the colored variants. If both top and bottom eyelids wore shadow, they seemed always to be contrasting colors. Lips, as now, were given a brighter red appearance, using a mercury-based (and hence, again, poisonous) plant dye called fucus.

The Egyptians were by no means alone in making use of makeup. India has as strong a tradition of painting the face as Egypt. Closer to home (for the Egyptians), all the major civilizations of the ancient world seemed to have made use of cosmetics to a greater or lesser extent. Whether the users were Babylonians, Sumerians, Assyrians, or Medes, both makeup and perfume were popularly employed, at least by the wealthy, and often wigs (tightly curled as opposed to the stunningly severe straight lines of the Egyptian equivalents) were used to add to the impact.

The Hebrews also made use of cosmetics. Though the Bible doesn't say much about this, it is notable that there is no direct prohibition of makeup. Admittedly, the only explicit mention is in 2 Kings, describing the despicable Jezebel painting her face, but it does seem likely that the features described positively in the Song of Solomon (cheeks like pomegranates and lips like a scarlet thread) were helped out by a pot of rouge. Perfume is mentioned much more frequently—again, that striking Old Testament love poem refers to "the scent of your perfume," while in the New Testament, in John's gospel, we hear of a woman called Mary bringing a pound of very expensive perfume, oil of nard, and anointing Jesus' feet with it. (The word "nard" was sometimes used to refer to lavender, but this was probably true nard, an aromatic oil derived from the root of the spikenard plant.) Even in the strictures of the epistles against women being showy, where they are told not to braid their hair or wear gold or pearls or to wear expensive clothes, there is no explicit ban on makeup or perfume, though later church fathers would rail against any use of paint, hair decoration, or adornment as unnecessary and very probably sinful.

The use of makeup continued unabated from the older Middle Eastern civilizations through their spiritual successors the Greeks and Romans, though the early Greeks seem to have made less use of cosmetics and their increasing employment in later Greek culture may have been due to exposure to other civilizations. Like the Babylonians, Greek women seem to have used a face makeup that contained lead sulfide to provide a virgin whiteness. It was probably realized at the time that this wasn't good for the complexion, but the slow poisoning from lead entering the bloodstream was not understood—lead was still being used in water pipes as late as Victorian times, and consequently this form of face paint, known as white ceruse, would cause illness and death all the way from ancient times to the nineteenth century.

One specialty of Greek cosmetics was an enthusiasm for blond hair that would be carried through to the Roman culture, leading to the famous comment of Pope Gregory the Great on seeing fair-haired slaves from the Anglian kingdom of Britain in the marketplace in Rome, *non Angli sed angeli*—not Angles but angels. Both Greek men and women used a version of the "sun-in" type of hair bleach, washing their hair with a special ointment, then sitting in the sun for an hour or more while the hair lightened in hue.

The Romans followed closely in the wake of the Greeks, picking up ideas from them and from the Egyptians on their conquests. Again, lead-based pigments were very popular, and adding to the risk of appearing beautiful (and emphasizing the lengths women were prepared to go to), it was not unheard of for Roman women to try to improve their skin with a face pack made from crocodile dung. We tend to think of beautifying as primarily a female obsession with the macho Romans, but given the number of gibes in Latin poems against men who dyed their hair or used facial makeup, it clearly must not have been uncommon among the patrician classes.

Romans also made significant use of perfumes, though the need was rather less strong with the Roman moneyed classes than for many civilizations before and after them, thanks to their enthusiasm for baths, not just as a means of cleanliness but also as a social gathering place. Baths were usually visited twice a day, and it was not uncommon for men to spend the whole day there, using the baths as a kind of social club for mixing and talking with the in crowd.

As Rome's influence collapsed and the West became increasingly Christian there was an almost entire collapse of the manufacture of perfumes, which like Greek philosophy would filter back to Europe via the Arab world. Makeup continued to a lesser extent, though it was to be much more subtle than had been common in pre-Christian times for a good number of years. By the 1500s, though, makeup had undergone a thorough resurgence—and still that white lead-based skin color was popular and was making women ill. On the whole, men had dropped out of wearing makeup, and those who still wore it were often scorned.

What would follow would be an ebb and flow of use of makeup that went along with fashion and was accompanied by a rich variety of hairstyles (and in some periods the use of full or even ludicrously large wigs). The enthusiasm for makeup continues through to the present day. We may have lost the deadly white lead, but the cosmetics market is immense, and the message is constantly pounded from the TV and the movies that no woman should be seen without appropriate makeup (and no man without the right, if more subtle, cosmetics).

All along, makeup has had the mixed allure of attracting the opposite sex and (for the older wearer) appearing younger. Gray hair was certainly being colored back in Roman times, but there is a much stronger pressure now on sustaining the appearance of youth. The latest antiwrinkle cream (with some appropriate near-science or pseudoscientific explanation of how it works) promises more and more each year, though the biggest adaptation to fight the appearance of aging would inevitably be cosmetic surgery.

Technically, piercing, tattooing, and scarring are all forms of cosmetic surgery, but they are examples of cosmetic body modification that blatantly state "this is not natural." In many cultures, the whole idea of these modifications is to make it obvious that the wearer is *not* part of nature—they are seen as a marker that the human being is more than an animal and anyone without them is considered subhuman. However, the way the term "cosmetic surgery" is widely used implies the entire opposite—a change that is designed to look natural. An improvement (depending on your viewpoint) over the original appearance, obtained by surgical means, that looks as natural as possible.

Cosmetic surgery itself is part of the wider scope of plastic surgery. Some of the enhancements that are made surgically are reconstructive—for example, after an accident has left damage to the appearance—but many others have exactly the same function as an externally applied cosmetic, making the patient look more attractive by providing an actual change to the body, rather than a veil for reality. With surgery the reward is usually higher—but so is the risk.

Like the other branches of cosmetics, surgery is big business. Nearly 11 million cosmetic surgical procedures were undertaken in 2006 in the United States alone. Broadly, the functions of cosmetic surgery are to change the shape or size of part of the body, to remove fat, and to remove external skin variations from acne to wrinkles that are considered unattractive. Sometimes the approach can be subtle—for example, making a face more symmetrical to increase its attractiveness—at other times the change is as simple as adding an implant, effectively internal padding, or involves making a feature larger or sucking out fat as if using a vacuum cleaner around the house.

It's worth spending a moment on that symmetry aspect of beauty, as it is very revealing about the whole nature of human attractiveness. There are many variants around the world and in different points in history on what is accepted as true beauty. Women, for example, have been considered beautiful when slim or fat, tattooed or clear skinned, pale of complexion or tanned, with small noses or large ones, and so on, through an endless and bewildering series of contradictions. But facial symmetry (along with youthful characteristics) is a prime marker of why a man feels that a woman is beautiful or a woman thinks a man is attractive in looks.

This is natural selection at work. The outward signs of many diseases produce a degree of asymmetry—when we see a face that has perfect balance, we assume health. This enthusiasm for symmetry has been tested using photographs where half faces are mirrored—the symmetrical version is nearly always considered more beautiful than the original face. When we look for a mate, at the basic biological level we are seeking out the ability to breed and looking for lasting health. The appeal of youthful characteristics is to increase the chances of being fertile (this also explains why women tend to be less worried about youth in a partner than men, for whom age is less significant to the ability to breed), while the symmetry, attractive in both sexes, is a marker of health.

It's easy to think of plastic surgery (whatever the application) as a modern phenomenon—and this certainly is true for the widespread use of the technique that results in 11 million cosmetic procedures in a year—but it dates back a surprisingly long way. Early examples (the earliest known twenty-seven hundred years ago in India) were confined to surgery to repair defects, rather than to improve the appearance. This wasn't due to any lack of interest in looking more attractive than nature allowed—the prevalence of cosmetics throughout the ages, despite regular rants against them by religious leaders and others, demonstrates that the market was there—instead it reflects the sheer danger and unpleasantness of operations before anesthetics, precautions against infection, and modern surgical tools and techniques were developed.

Modern cosmetic surgery was a spin-off from the developments in reconstructive plastic surgery that were made necessary by the horrifying casualties of all-out war. World War I particularly resulted in many facial injuries that needed reconstruction. Although the techniques of the time were crude, as they developed and infection control became the norm, it was inevitable that surgeons began to realize what a big market there was for applying these reconstructive skills to making improvements to an uninjured patient.

Reconstructive plastic surgery became better established as a discipline in the 1930s, but it wasn't really until the 1970s that cosmetic surgery gained a legitimacy that made it acceptable to a wider audience. This doesn't mean that cosmetic surgery hadn't been performed earlier. Once we had moved from the constricting ethos of the early nineteenth century into more of a consumer culture, it was inevitable that there would be a resurgence in the cosmetic arts, and initially many who offered some kind of cosmetic surgery were quacks who were able to do much more harm than good. A classic example was the use of paraffin injections, employed to even out wrinkles. The effect was good initially, but the paraffin had a distinct tendency to sag in the heat and not infrequently would cause cancers.

Many of the basic techniques used today were undertaken

(with mixed success), by Dr. Charles C. Miller, of Chicago, whose 1907 textbook, *Cosmetic Surgery: The Correction of Featural Imperfections,* included face-lifts and nose jobs. Miller seems to have straddled the borderline between quacks and true cosmetic surgery—his reputation was not great, yet he seems to have had some successes. Even today, there is a mixed feeling about those who perform cosmetic surgery, that somehow they pervert their medical knowledge by applying it to a frivolous task. Cosmetic surgery is certainly much more socially acceptable now than it was in the midtwentieth century and is much safer than it used to be. Even so, there is risk attached to *any* surgical procedure, in terms of both the operation failing to improve the patient's looks and the risk of complications up to and including fatalities. Cosmetic surgery may be an essential in the world of celebrity but is still regarded with distaste by many members of the general population.

For those for whom going under the knife seems too extreme but simply masking with makeup isn't enough to achieve that special look, it is possible to have forms of makeup that do more than act as a simple colorant or concealment but instead make a direct change to the body itself. Perhaps the best-known early example of this was belladonna. Also known as deadly nightshade, this is a plant of the same family as potatoes, tomatoes, and eggplants that, as its alternative name suggests, contains a poison—one of the strongest alkaloid toxins of any plant in the Western Hemisphere. This was a poison that was used to change the appearance.

When a solution of belladonna extract was dropped into the eyes, the result was a partial paralysis of the iris, resulting in pupils that dilated—the black central circle of the eye became bigger than usual. (Quite how this was ever discovered, like the discovery that cooked kidney beans are edible [see chapter 2], is hard to imagine. Who would say, "Let's drip this deadly poisonous plant extract into our eyes and see if it makes us look prettier"?) Using belladonna wasn't very helpful in terms of eyesight—the whole point of the pupil is that it cuts out light when it's too bright, so partially stopping it working means the user finds light dazzlingly brilliant and has a degree of blurred vision. But it's true that belladonna did have a noticeable effect on attractiveness.

An interesting experiment was carried out in the 1970s. Apparently identical photographs of a woman were showed to men. Almost uniformly, they decided that one of the pictures was more attractive than the other. The only difference between the two was that one photograph had been modified to make the pupils larger. We find members of the opposite sex more attractive when their pupils are dilated. This seems to be because our pupils naturally dilate when we are sexually aroused. Seeing someone of the opposite sex with dilated pupils gives a feeling of a more open channel of communication to his or her inner being and suggests mutual attraction. It is true that the belladonna users were putting their eyesight at risk, but the cosmetic function was more than an old wives' tale—it really did (if only marginally) increase the attractiveness of the user.

It might seem quite incredible now that our ancestors were prepared to drip a neurotoxin into their eyes to make them look marginally more attractive. It shouldn't be, though, because modern beauty seekers go one step further and have an even more powerful poison injected into their faces. The botulinum toxin, common in poorly canned foods and meat products before modern preservation methods, is an immensely powerful poison that requires only a thousandth of a gram to kill. It is produced by a bacterium, *Clostridium botulinum.*

Botulinum is a potent neurotoxin that results in weakness, vomiting, and paralysis, followed by total failure of the respiratory system. Hardly a substance you would expect to be used at parties and in shopping malls—yet that has been the remarkable fate of products like Botox and Dysport, small injected doses of the botulinum toxin, initially introduced for treatment of debilitating muscular conditions but soon also used to smooth out facial wrinkles.

The amount of the toxin in a Botox or Dysport injection is very small, yet even so it is remarkable how easily these products have become accepted, considering what it is that they do. They work by effectively reducing the strength of the underlying muscles by disrupting the nerve signals to them, reducing wrinkles that are produced by musculature at the cost of an artificial, ironed-out-looking skin and in some cases a loss of ability to display full facial expression. The wrinkle removal is not permanent—the treatment gradually wears off after a few months.

Like cosmetic surgery, the use of Botox and Dysport is a form of intrusive body modification. Body modification, which encompasses tattooing, has a much wider scope and has been practiced in many forms over the years. Beyond tattoos are piercing, cutting, binding, implanting, forced growth patterns, and, arguably, aspects of body weight and changes of musculature that result from exercise or dieting. No one can look at a bodybuilder and think that this is a natural human figure—whether you like the appearance of bulging biceps or not, it is a significant update to the normal muscle structure.

Bodybuilding has been around since the time of the ancient Greeks, although the aim at the time was to diet and exercise to improve performance in the games rather than any concentration on physique for its own sake. The use of stone clubs or dumb-bells called *nals* in India seems to date back up to a millennium and also had more of a focus on sporting and physical prowess than sheer looks. It wasn't until 1899 that the modern concept of bodybuilding for its own sake, showing the body off almost as an art form, was devised and publicized by the German strongman Eugene Sandow. Today, of course, it's big business and has even started one career that ended up with a state governorship.

An awareness of the impact of diet, like so many other actions to change our appearance, goes back a long way, too. If diets were rarely mentioned in any detail in earlier times, it was likely to be more a matter of lack of choice than a failure to realize that what you eat influences how you look. For hundreds of thousands of years, like most other animals, humans ate what they could find and the whole business of hunting and gathering food provided enough exercise to balance out what was eaten. On the whole, food was in scarcer supply than dietary needs. With the organization and structures of civilization, though, this would change.

Ancient advice on diet is of variable value. Roger Bacon commented that "[a] real remedy . . . might be found if a man from his youth would exercise a complete regulation of food and drink . . . ," going on to list other contributory factors but not telling us what such a complete regulation would consist of. We know that a reduced-calorie diet was encouraged by the Venetian Luigi Cornaro in the fifteenth century, though like Bacon his intention was more to prolong life than to improve looks—and a extremely low–calorie diet has indeed been shown to provide some degree of life extension (see page 58), even if many would argue that the benefits are outweighed by the lack of enjoyable food.

It was not until the nineteenth century that any concerted effort was made to change diet with the specific intention of improving appearance. There was obviously some word-of-mouth expectation of how what we eat influences our looks, as when in 1860s England a man by the name of William Banting tried to reduce his extreme obesity. He was recommended to cut back severely on sugar and starch, resulting in a weight loss of fifty pounds in one year. Others would chip in with their own ideas, notably Dr. John Harvey Kellogg, whose enthusiasm for fiber resulted in his invention of the cornflake (though this was more with an eye on regular bowel movements than the body beautiful). Yet there was no real scientific idea of a direct comparison of the content of food and its impact on the body until 1918.

This was the year that the modern diet was born with Lulu Hunt Peters's bestselling book *Diet and Health, with Key to the Calories*. Diet books would regularly appear on the bestseller lists from then on. Over the twenty years or so before Peters wrote her book, information had been collated on the energy content of foods. (Confusingly, the modern scientific measure of energy content is joules, but the old measure, calories, still tends to be used in diet. Even more confusingly, what is usually referred to as a "calorie" is actually a kilocalorie—a thousand calories—but at the time the measure was popularized it was thought that the general public couldn't cope with anything so complex as a kilo-anything. It is sometimes distinguished from a "real" calorie by giving it a capital *C*, but this is not consistently used.)

By measuring the energy content of the food, it was relatively easy to see what was necessary to keep the body going and what provided an excess that would contribute to fat. Although the body's metabolism is a complex mechanism, especially when other nutrients are taken into account, in terms of dealing with calories and fat the equation is simple. If a dieter ate fewer calories than he or she consumed in exercise, then the body would have to make up the excess from its natural store—fat.

After Peters's book came a flood of diets, each suggesting a totally different option to lose the pounds. Arguably a strong influence on the urge to diet was the introduction of photographs of beautiful people in magazines and the movies, showing an idealized form to aspire to. From the point of view of the authors of diet books, a cynic might point out that it didn't really matter if the diet worked—as soon as a new diet arrived, thousands or millions of desperate people would give it a try and send money rolling in. This was all too often a literary version of quack medicine. Certainly many diets were based more on a top-of-the-head idea than any scientific basis. Readers were presented with the Hollywood (grapefruit) diet, diets based on different food combinations, diets involving seaweed, hot water, low carbs, fruit, and much more.

Diet has become even more of an issue in the last fifty years. With so many more people in the West overweight—many now clinically obese—dieting to lose weight and to change body shape has become an obsession. The sheer easy availability of fats and sugars, substances that are naturally relatively rare, so we tend to eat all we can of them when we come across them, is leading us astray on a huge scale.

Unfortunately, there are several problems with dieting to lose weight. One is that diet alone doesn't have the ideal effect. It has to be combined with exercise, which not every dieter is willing to undertake. Another issue is the lack of lasting effect. We like to think of a diet as a short-term change, to lose a few pounds, after which we go back to our old ways. In reality, diets can only be truly effective if kept up indefinitely, a fact that has made dieting organizations such as Weight Watchers very profitable. Then there's the lack of appealing low-calorie food, though this is to some extent being addressed.

Those whose careers depend on their appearance show just how much is possible, losing weight until they become stick thin, but this requires dedication that only the driven can sustain and often produces feelings of depression and a highly unattractive regime. There is also a distinct possibility that by using these extreme diets individuals put their long-term health at risk, trading short-term appearance for premature aging. Dieting is something we can all play with to make minor modifications to our bodies—and a good, balanced diet combined with appropriate exercise is the ideal we should all aspire to—but when dieting takes over your life, it has gone too far.

For much of the twentieth century the more explicit and direct forms of body modification, unlike dieting, were relatively unpopular in the West—and some of them still cause revulsion in more traditional circles—but since the 1970s there has been a surge of activity in this field, particularly around piercing. Where traditionally body modification has been a matter either of tribal conformity—as, for instance, in the use of body paint in ancient Britain—or of being more attractive, the new wave of body modification combines the tribal aspect (being like other punks or whatever group membership is being displayed) with a more inward-looking self-confidence factor.

The implication of having a modification is that the individual has control over his or her body. (This is a direct parallel with the attitude in many tribes that practice body modification like scarring. As we have seen, they consider the fact that they are modified shows that they are apart from and above nature. They are in control, unlike the animals.) As one body modifier commented, "I make a statement, I've chosen myself, I am part of a culture but I don't believe in it. My body modifications are a way to say that." Of course, in reality, practically all body modification shows strong attempts to identify with a particular culture rather than stress individuality, but it's a chosen subset of the culture, not the image defined by the media, your parents, or the government.

The key themes of body modification remain much the same now as they were in ancient times—attractiveness to the opposite sex (or to the self in boosting self-image) and reinforcement of tribal membership. In modern times, there is a clear distinction between light tattooing and piercing—one or two earrings or a small, subtly placed tattoo, for instance—and more heavyweight modification. The (more common) light modification is primarily about attraction. As sociologist Paul Sweetman observes, such modification is often regarded "in primarily decorative terms: 'I thought it would look nice,' or words to that effect, being a common response to enquiries regarding their motivation . . ." It's very similar to wearing other forms of jewelry—light modification is an adornment.

Heavier modification tends to be more closely associated with tribal membership, although it can also have the aspect of a statement of personal identity in our historically unusual culture that is obsessed with the self. A recurring theme in interviews with people who have modification is that they see it as a way in which they can be made different, are able to stand out from the crowd—though this is rather ironic when so many people now have some form of piercing or tattoo and, arguably, among adults it is now those with no modification at all who stand out.

A more global awareness has also increased the interest in body modification inspired by different cultures. Where, for example, tattooing used to be very much the mark of certain fields of work (sailors, for instance) in the West, the wider spread of tattoos has seen influences come in from farther afield, adding an exotic flavor to fashionable appearance. Fashion here seems to include a machismo desire to have an adornment that the wearer is seen to suffer in order to receive. "Real" tattoos are considered quite different from temporary tattoos, which though visually similar are really an extension of the body painting tradition. Obtaining a real tattoo is a mark of passage, as much as any tribal ritual, even though the wearer may live to regret it, because modern fashions change much more quickly than tribal standards that last from generation to generation.

Alongside the increase in body modification since the 1970s has come a move in the West from a tradition of concealing the body (particularly the female) with heavy restrictive clothing to exposing more of the body and the body's shape with lighter clothes and a wider expanse of skin on show. Over hundreds of years just what was acceptable to display has varied wildly, but recent trends seem to emphasize the same "I'm my own person" statement as modification, as opposed to, for example, the medieval wimple, which was in part a mark of a woman's status as chattel.

With more skin on show, there is more pressure to work on the body itself through diet and exercise, to display the culturally accepted views of what is attractive. As we have seen, this is in direct conflict with the unhealthy diet prevalent in the West, a diet that leads to a much higher incidence of obesity, resulting in a particularly high disparity between what is accepted as attractive (at the extreme in women, the size 0, stick-thin fashion model) and the appearance of the typical individual.

The enthusiasm for being thin that is dramatically underscored by near-anorexic celebrities is a highly localized fashion in time and culture. The Western obsession with thinness is nothing like universal—in some African and South American countries (as also seems from the statues that remain to have been the case in prehistory), being fat is seen as an essential for attractiveness, probably reflecting the fact that only those who were particularly favored could have enough to eat to be overweight (where now it is often the poor who are overweight). In some of these cultures, young women are expected to stay in a "fattening house" on a special diet until they reach acceptably large proportions.

More extreme forms of body modification have been practiced in localized forms (localized in time or space, or both) in practically all parts of the world. We tend to think of African and Asian examples, like the frequent use of scar marks, producing raised patterns on the skin as a sort of bas-relief equivalent of a tattoo, but it ought to be remembered that in nineteenth-century America and Europe a drastic form of body modification was regularly practiced. An unnaturally thin "wasp waist" was considered attractive, when as a result of compression of the waist using corsets and girdles a normal measurement of perhaps twenty-eight inches was reduced to as little as twelve inches.

However, we shouldn't ignore other dramatic modifications. The so-called giraffe women of the Padaung (or Kayan) people come from the mountains between Burma and Thailand and would resemble other women of the region were it not for their unique appearance caused by the brass coils they wear around their necks. These are fitted from around the age of five, with the wearer upgrading to longer and longer coils over time. (It is often thought that they add extra rings as they grow older to increase the extension, but the coils are a single piece of metal.) The "giraffe" label is misleading—although the women look as if they have extra-long necks, it isn't possible to extend the vertebrae (the spinal cord would snap). Instead, the breastbone and upper ribs are pushed downward, giving the impression of a longer neck that is enhanced by the continuous line of the coil.

Like many other forms of modification, this is both designed for attraction—women with coils are considered beautiful in their culture—and a marker of membership, identifying the women as part of the tribe. The coils are a considerable burden— weighing around ten pounds and causing bruising to the skin—but don't impose the terrible price that legend attributes to them of causing the wearer to die if the coil is removed because the extended neck cannot support itself. There is no neck extension, and the coils are sometimes permanently removed.

A true extension that is reflected in some modern Western ear piercing is that applied to the lips both in Africa and in South America. Wooden plates (sometimes clay in the African examples) are inserted into the lower lip (and occasionally the upper lip, giving a ducklike profile). The lip is gradually stretched, starting with a small incision held open by a peg and working up to plates than can be as much as ten inches across. There can be no sense that this is a practical modification. It is inevitably a mark of status that in some cultures is restricted to the men or the women, in others used by both sexes.

These lip extensions, like the peacock's tail, are more than lacking in functional value—they are intensely impractical, restricting the natural functions of the mouth and lips and risking dangerous infection. Yet even they are not as restrictive as another form of modification, Chinese foot binding. This practice, now banned, involved wrapping a girl's feet in tight bandages from around the age of five or six. As the feet grew, they would be deformed by the binding, the toes breaking under the stress.

Women with bound feet were in constant discomfort and suffered many secondary problems. The result was a grown woman with feet that could be as short as three inches in length. This was regarded as an attribute of a beautiful woman, rather than producing a foot that was itself beautiful, though it is hard now to understand the appeal of such intentional malformation. It has been suggested that a more logical intent was to make the women unattractive to Tartar raiding parties, who would not be interested in women with such an alteration, but this seems an unlikely extreme measure to adopt.

Again this is body modification that went strongly against the practical—a woman with bound feet was unable to walk normally and unsuited for work in the fields (so the practice never caught on in the poorest families). Other societies have bound other parts of the body to change its shape—ancient Egyptians, for instance, bound the head while the bones of the skull were still changing shape to produce what was considered a more aesthetically pleasing shape.

Perhaps the most dramatic modifications are those that involve removal of part of the body. Many cultures have at one time practiced circumcision, though modern scientific thought generally regards this (particularly the female version) as nothing more than mutilation. It certainly has no practical value, unlike another modification that was common until relatively recent times—castration.

While there was a tradition of removing the testicles to produce servants—eunuchs—who could work in close confines with women (then regarded as property) without the risk of sexual interference, the subtlest application of this relatively simple though painful operation was the production of castrati: boys whose voices never broke and so were able to sing high parts throughout their working life. This was necessary for centuries when it was unacceptable for women to appear on the stage, leaving men to take their parts. Though those who have heard castrati sing emphasize the unique tone and power they had, few would now argue that a potential career (many of these operations failed to produce a great singer) was worth the price.

It's easy when thinking of modification for cosmetic appeal to concentrate on changes that we can see, but smell has also always had a complex role to play in the business of attracting the opposite sex. In the "natural" unwashed state, many find smells that build up around the body unattractive, and from early days different kinds of perfume have been worn both to mask body odor and to add an attractive scent, though ironically in later years, when bathing became more common, it was possible to lose too much of the natural odor, hence the famous request from

Napoléon to his wife, Josephine, saying that he would be home soon and requesting that she did not wash.

Fragrances were widely used in ancient Egypt, and like many societies since then that have been weak on hygiene, they frequently employed scents to cover the unpleasant smells coming from their bodies and their surroundings. Even today, in formal processions senior British judges carry a "nosegay," a posy of "sweet-smelling herbs" and flowers, originally held so they could bury their noses in it to cover the stink of the street, though now just adding to the ceremony of the occasion.

That there was a problem that needed addressing seems obvious from the lengths to which the rich of ancient Egypt would go to improve their body odor. Before important events and dinners, a great blob of ointment, strongly perfumed, would be placed on top of the inevitable wig. The ointment was sometimes contained in a cone-shaped vessel—which must have looked very odd. Body heat gradually melted the ointment, which then oozed through the wig, dripping onto head and body, providing cooling and a refreshing smell, if giving the wearer a most unpleasant sensation. The perfumes in such ointments seem to have been more herbal than floral.

The word "perfume" was not originally applied to a way of creating nice-smelling people. As a verb, "to perfume" meant "to fill something with smoke, to fumigate it," and from this "perfume" became a term for incense before settling on the kind of scents used to make one human more attractive (or at least acceptable) in smell to another. Perfumery is not only a very old business, but it has changed less over the years than practically any other. Although more artificial sources are now used, there is still a focus on distilling the essential oils from plants, making them more concentrated to give a pungent odor.

The only major breakthrough in perfumery between ancient times and modern was in the fourteenth century, when the Hungarians realized that by giving the perfume an alcohol base, rather than water, it would evaporate more easily, alcohol having a lower boiling point. This "Hungary water" resulted in the scents of the perfume being broadcast into the air more easily and as a side effect tended to preserve the complex chemicals better, giving the perfume a longer shelf life.

Traditionally, perfumes had two roles: to make the wearer smell attractive and also to mask less pleasant odors. At various points in history, the attitude to personal hygiene has varied considerably, but for much of the time baths were uncommon, certainly for the common folk, and even the aristocracy were not particularly worried about regular bathing. The answer was to cover the smell with perfume. Yet once it was more fashionable (and practical) to keep clean, it became obvious there were some body odors that even regular washing wasn't enough to deal with—particularly from the armpit, where a concentration of sweat glands leads to an easy buildup of stale sweat. The sweat itself doesn't have a bad odor, but the bacteria that grow on it produce unpleasant-smelling by-products.

The MUM deodorant seems to have been one of the first on the market back in 1888, a waxy cream, designed to kill the bacteria that caused the odor rather than reduce perspiration. Later formulations, usually incorporating aluminum chloride or chemicals with similar effects of blocking the sweat glands that produce moisture, would target the sweat itself. The unpleasant-to-use pastes were later joined by innovations like the roll-on in the 1950s (inspired, apparently, by Laszlo Biro's ballpoint pen) and aerosols in the 1960s.

Perhaps the biggest adaptation we have made to ourselves in the field of attracting others is to move away from the treadmill of the reproductive cycle. Our natural design is to be relatively briefly attractive during the peak of potential reproduction and after that appearance really doesn't matter. Once people had their families, they could let themselves go, dressing dowdily, doing little to improve their appearance. Only a small proportion of the population—the rich and the highborn—would worry too much about trying to continue attractiveness past the thirties.

Though apparently trivial, it is probably one of our biggest rebellions against the basic human design that now practically everyone makes an effort to extend their attractiveness into a period of their life when—for biological reproductive reasons—it's unnecessary. This is apparent in attitude to clothing and hairstyle and in the use of beauty and antiaging products, but nowhere is it more obvious than in the role of cosmetic surgery.

As we have seen, some cosmetic surgery is performed on younger people, when their jobs require visual perfection (in the movies, for example) or to achieve some personal ideal of beauty, but the real driving force behind this technology is the attempt to undo and hold back the changes that aging and often an unhealthy lifestyle have made to the body. A spot of liposuction, a nip and a tuck, and you're looking ten years younger. That so many people are willing to undergo potentially painful procedures, always with some attendant risk, emphasizes just how important this need to stay visually young is to us now.

Where once almost all the cosmetic business was focused on

young people, helping them to attract the right mate, now it is a major prop in our denial of aging. To the extent that makeup and even surgery changes the way we regard ourselves and the way we act, keeping us youthful longer, it is a force to be reckoned with in our adaptation from Human 1.0.

The gene therapy that we've already seen as a possibility for extending longevity also has the potential to make extremes of personal appearance possible. It's now relatively common for genetic engineers to produce creatures that glow in ultraviolet light. The jellyfish *Aequorea victoria* has a natural green glow when irradiated with ultraviolet. Technically the jellyfish is fluorescent, not luminous. Fluorescence is the process of absorbing one frequency of light (typically invisible) and giving off a different frequency (visible), giving the appearance that the fluorescent object is glowing, though in fact it's not generating light in its own right, like the truly luminous firefly, but is rather shifting the color of the light it reemits.

There is a protein in the jellyfish that absorbs ultraviolet and gives off green light. By introducing the gene that produces this protein into other creatures they, too, can be made to glow in ultraviolet. This has been done with mice, rabbits, fish, and more. The scientists weren't introducing the gene to produce glow-in-the-dark freaks but rather as markers for genetic change, making it easier to see how cells were influenced. Yet it didn't take long before one of the products—glowing zebra fish—was being sold commercially.

The National Institute of Environmental Health Services has been funding research into an even more dramatic version of the glowing zebra fish. They are trying to introduce the gene that produces luciferase, an enzyme produced by fireflies. When mixed with a compound called luciferin it results in a true glow. The idea is to give these fish genetic modifications that only enable luciferase production in the presence of pollutants, making the fish natural warning lights of unwanted chemicals in the water. Another team from the University of Hertfordshire in the United Kingdom has suggested introducing luciferase-generating enzymes to Douglas fir trees to make self-illuminating Christmas trees.

All these glowing creatures and plants don't provide any direct parallel for genetic cosmetic effects for humans, as they involve germ line modification. The genes are inserted into the newly formed embryo and become part of the entire animal's or plant's DNA. This means they are with it for its entire life and will be passed on to future generations. The dangers of this level of modification are too extreme for cosmetic effects in humans. What may hold out more hope for the future is the ability to perform this kind of gene modification temporarily and locally to give, for instance, glowing hands—but as yet such an idea is pure science fiction.

More practical is the idea of video tattoos—moving, potentially glowing images on the skin. These have a powerful appeal to many people, even some who might not consider wearing a conventional tattoo. A temporary version has been dreamed up by futurologists at the United Kingdom's telecom company BT, which could be available in five to ten years' time. The basis for a stick-on temporary video tattoo is a film-based display.

For many years electronic firms have been working on electronic paper—a computer display, probably based on a variant of the liquid crystal, that is as flexible and easy to roll up as paper.

Industry giant Philips has already prototyped a millimeter-thick display that is flexible enough to wrap around an arm or leg. BT is predicting that the first video tattoos will be thin sheets of polymer film that are stuck onto the skin, where they can stay in place for several days. The e-tattoo could display preset videos but also act as a cell phone or be Internet connected.

Further out, such video tattoos could be embedded in the skin, based on nanotechnology. Instead of applying a film, each pixel on the display would be a separate nanodevice. These would be sprayed onto the skin and would locate themselves with respect to the other devices, setting up a matrix of tiny display points that come together to make a picture. Variants could be sticky and instead of embedding could rest on a surface, from a brick wall to individual hairs, providing hair coloring that could be changed with mood (or could flow with different colors).

Such a vision of the future is highly speculative, but what we do know from the past is how far we are prepared to go to develop cosmetic technologies. The market is so big and the pressure to produce something new and even more impressive so great that there are bound to be developments of these sorts as soon as the technology is available to support it. We might feel that all the great scientific breakthroughs should be made for life-saving medical reasons or to understand the basics of physics, but realistically there are more dollars available for cosmetic applications than (say) finding fundamental particles.

Attracting a mate is an essential for survival of the species, but it isn't enough in itself. As humankind has demonstrated all too well in our ability to wipe out other creatures, in the face of a dangerous predator even a fecund species can be destroyed. Defense and the collection of food proved a forceful driver for change. It was inevitable that the upgrade process would include ways of enhancing our natural strength, that we would develop approaches that enabled us to go beyond the basic human physical capability. By nature we are more 120-pound weakling than powerhouse. But we wouldn't leave things to nature for long.

4.

The Strength of Ten

O! it is excellent to have a giant's strength, but it is tyrannous to use it like a giant.

—William Shakespeare, *Measure for Measure*

A human fist can do a certain amount of damage, but a fist holding a rock or a club can do much more. Weapons and manual tools soon extended our natural capabilities. Even the crudest technology gave the upgraded human a huge advantage. The difference in strength between two different prehuman species can be measured in a few percentage points. By using weapons, the strength of the enhanced human was magnified many times over.

It all began with sticks and stones, a prehistory of violence unconsciously reflected in the child's rhyme "sticks and stones may break my bones, but words will never hurt me." Unworked stones initially increased the power of the fist or were thrown as a simple barrage. This is a classic example of how easy it is to miss the power of upgrade. We can't get excited about a stone as a piece of technology. Stones lie around on the ground. No work is required produce them—all we've got to do is pick them up and use them. Yet in the hand (or even more so with a low-tech slingshot like the one David used to take down Goliath) a thrown stone vastly increases a human being's reach, and a pounding stone transforms the power and danger of a punch.

Weapons that are enhancements of the handheld rock continue all the way up to the present day. Many of these developments make the "rock" longer to provide greater reach and leverage, typical in that most common of weapons, the club, which can be anything from a crudely hacked off tree branch to a modern high-tech baseball bat. Extra sophistication was added to this development, both in providing iron tips and including features like the "morning star," a spiked ball on a chain attached to the end of the club.

As projectiles were being made more sophisticated with addition of the simple slingshot to the stone, so the second generation of weapons emerged based on a sharpened rock or a pointed stick. The arms race began early in our prehistory. Before long, weapons were being given an edge by chipping away at stones like flint to produce sharp cutting blades by breaking the rock along flat fracture planes. Sharp blades were also thrown from early times. Darts have been found dating back around eighteen thousand years. These are not like the modern conception of a blow dart or a hand-thrown sports dart but were more like short stone-headed spears with some form of fletching (feathers or similar materials to help them fly straight) that were launched with an atlatl, a handheld throwing device that cups the dart and flings it, giving it sufficient force to embed itself well into a wooden plank (or a human torso).

A well-made flint knife was very sharp when first chipped but wouldn't retain that sharpness for long. The magic of metal, first coming into significant use in the weapons of the Bronze Age, was that it combined the hardness of stone with a lasting sharpness and an ease of working into complex shapes. First bronze (a hard alloy of the softer metals copper and tin) and then iron made stone weapons largely redundant.

Over the ages, swords themselves also evolved. Based on the pointed stick, these were originally weapons for jabbing and impaling. By Roman times, the edges were adapted for slashing, producing a double-use weapon. This was still the case at the time of the Norman Conquest, but later in medieval times the edges of the sword, particularly for the foot soldier, became more important and the most dramatic form of the sword, the two-handed great sword, which could be up to six feet long, came into play, used in a scything action to blast through the opposing troops.

By the Middle Ages, the advantages of projectile weapons—the ability to kill from a distance or when behind the protecting walls of a castle—became obvious. Bows had been in use for many years. The earliest examples that have survived date back around ten thousand years, and bows could have existed even earlier. Ötzi, the fifty-three-hundred-year-old man found frozen in the ice (see page 83) had an unfinished yew bow with him. But bows were perfected in the fourteenth century with the longbow, around six feet of yew providing a powerful driving force for the arrow. Many English churchyards still have yew trees planted in them, and though it's possible that some of these were there as a carryover from pagan traditions, they were certainly encouraged by British royalty as a source of weapons for their troops.

Slower to reload but even more deadly in penetrating power was the crossbow, which first arrived on the military scene in the twelfth century. Like nuclear weapons in the twentieth century, the crossbow was considered so dangerous that it was thought to give an unfair advantage, and a number of popes announced the equivalent of a crossbow proliferation treaty, banning it from the battlefield (though Pope Innocent II did stretch the point so that it could be used against the infidel in the Crusades).

Guns, bombs, and all the explosive might that would bring modern weaponry into being had, as we have seen when looking at armor, already existed in parallel with the more conventional arms of medieval warfare. Gunpowder, the first military explosive and propellant for guns, was recorded as early as A.D. 300 in China, and it's likely that the idea was exported from China to the West. We don't know this for certain, but this black powder (made from a mix of carbon, sulfur, and saltpeter) seems to have emerged fully formed as an explosive in the thirteenth century. Roger Bacon, the English friar and early scientist, has often been described as the Western inventor of gunpowder because his description of it in his letter "Concerning the Marvelous Power of Art and Nature, and Concerning the Nullity of Magic" is the earliest clear written description of black powder, but there is no reason to think that Bacon invented it.

In practice, Bacon's formula isn't very effective. The proportions he gives would produce a powder that was more useful for fireworks than a gun—but this reflects the origins of gunpowder in pyrotechnics. Bacon does describe the use of gunpowder in warfare:

For the sound of thunder may be artificially produced in the air with greater resulting horror than if it had been produced by natural causes. A moderate amount of proper material, of the size of a thumb, will make a horrible sound and violent coruscation. Such material may be used in a number of ways, as, for instance, in a case similar to that in which a whole army and city were destroyed by means of the strategy of Gideon, who, with broken jugs and torches, and with fire leaping forth with ineffable thunder, routed the army of the Midianites with three hundred men.

Initially, true gunpowder weapons were limited to cannons, both in China (from around A.D. 1000) and in Europe in the early fourteenth century. The indifferent quality of early gunpowder, combined with the need to use a big vessel like a cannon, made these original gunpowder weapons near-impossible to manage as offensive weapons. They were as likely to be employed to establish fear in those attacked as to do direct damage. It wasn't until the early fifteenth century that small handheld "cannons" were developed. For two centuries the main form was the arquebus, a smooth-bored long gun. This looked like a rifle but can't be called one as it didn't have the spiral cuts inside the barrel—rifling—to spin the bullet for more accuracy of flight. It was its successor, the musket, that finally received that rifling and retired plate armor from the battlefield.

The range of arms and armory that humanity has produced over the ages is so large that it would take a book much larger than this one to explore just this one aspect of humanity's vast upgrade project. It's doubtful whether weaponry is humanity's most noble addition to our natural capabilities, particularly when used against our own kind, yet it is certainly true that in pitting ourselves against the dangerous predators of our world the development of arms has proved the decisive factor in our survival.

Sadly, it has also led to the extinction of some of the creatures that have come up against us.

Much weaponry since the cannon has ceased to be a true extension of the body. It's just about possible to argue that a slingshot, or even a longbow or gun, is an extension of the body's ability to pick up a stone or a stick and throw it, but everything from the cannon to the cruise missile is semiautonomous, not so much an extension of the human being as an extra contributor to the battle.

I will return to the impact of technology later in the chapter, but adding a weapon to the outside of the body was not to prove the only means of enhancing our muscle power. As we have already seen (see page 105), building the muscles for athletic and military reasons goes back a long way, whether through exercise or diet. More recently, chemical means have been added to ways we can work on sheer strength.

At one point the urge to provide enhanced strength seemed to rely more on a version of sympathetic magic than science, injecting extracts from powerful creatures. This seemed to be the case, for example, in the use by the British physiologist of American and French extraction Charles-Edouard Brown-Sequard of extracts from animal testicles to try to improve body strength and longevity. But more scientific treatments came in with anabolic steroids, which were first used in 1930s and became common in professional sports by the 1970s. Other performance enhancers, like erythropoietin (EPO), which increases the red blood cell count and so increases the blood's ability to carry oxygen, have since been added to the illegal sports drugs that have been used to beat the best.

Now we can go further and add genetic modification to the

methods available to enhance strength. There is no doubt that once it is possible to make safe and predictable genetic modification (see page 53), it can be used to "design" babies who will be, for example, better atheletes, if this is a desirable thing to do. At the moment we tend to regard "built-in" and long-term modifications, like a genetic benefit or a healthy upbringing, as acceptable contributors to athletic prowess, while short-term "easy-fix" solutions like performance-enhancing drugs are not acceptable.

Author Bill McKibben argues that making genetic modifications to improve the potential to be a sportsperson is one step too far. "If that happens," he asks, "what will be the point of running?" But McKibben misses the important issue here, which isn't the unfairness of having genetically modified athletes but rather the pointlessness of much competitive sport itself, which is based on a totally arbitrary concept.

The way we undertake competitive sport is not designed to show how well individuals push themselves against their basic capabilities, how hard they have worked, what stars they are—instead our sporting ability largely depends on genetic input. It wouldn't matter how much effort I put into making myself a runner; I would never be able to compete effectively at a county level, let alone in national or international competitions. It's the luck of the draw. I'm not a runner, and I never will be. (Or a soccer player or a baseball player, or a serious competitor in anything more physical than backgammon, for that matter.) If we really wanted a "fair" race, everyone would be handicapped according to their natural ability and only what they achieved through extra effort would be rewarded—but of course that's never going to happen.

What McKibben is saying is that it is okay to have an athlete's ability decided by random factors but not by planned effort on the part of his or her parents. This might be right, but it's still the nature of competitive sports that causes the problem, and rather than worry about this genetic aspect, it would be better to find some method of handicapping to natural ability. Otherwise just where do you draw the line? Should sporting champions not be allowed to have children together, because their genetic makeup is liable to result in better-than-average children when it comes to sporting ability? Hardly. If it were possible to safely and routinely make a genetic modification that resulted in better physical capabilities and there weren't any side effects (such as depressed mental capabilities or reduced life span), it's not so much a change we should be attempting to suppress as one we should be attempting to make universal.

Yet even this apparently egalitarian approach of enhancement for everyone doesn't suit McKibben. He imagines an enhanced runner—nothing special, just an ordinary guy in a marathon—who because of his enhancement doesn't have to go through all the suffering and training that current marathon runners (including McKibben, strangely enough) put themselves through. Again, this misses the point. If human beings are enhanced, we shouldn't be worried about how they fit into the totally arbitrary races we now have. (The marathon is the ultimate example of arbitrariness—the bizarre distance of 26 miles and 385 yards doesn't reflect the original run in Greek history it is supposed to honor. The 385 yards segment was added in 1908 when the Olympic race was run in Windsor, England, at the request of the royal family to bring the start into view of the royal nursery, and the race has stayed that length ever since.) Sports should be designed around the people who take part in them, rather than the other way around.

To the enhanced person, a marathon may be the equivalent of a fun run for an ordinary person (something I find challenging enough)—but so what? Should we ban fun runs because you don't have to push yourself to ridiculous limits? There is something very sad about the obsession with doing things because they stretch human beings to the limit. This is what leads to attempts to trek across the North or South Pole on foot or to climb Everest without oxygen. To any sensible onlooker, it's stupidity. There is no scientific benefit. There is no discovery. It's little more than risky posturing.

The ultimate example of this madness is that awards have already been funded for the first person to climb Mount Olympus on Mars, to cross the Martian poles "without airborne support or resupply," and to descend the vast Valles Marineris on Mars "using no technological support other than that required for life support and basic mountaineering." Leaving aside the total strangeness of these challenges (for example, what airborne support do they have in mind? There is no air on Mars), this is Boy's Own stuff that now seems hugely dated—it is celebrating vast effort for no benefit whatsoever. You might as well have an award for the first person to hop all the way around the Moon or the first person to eat a whole asteroid (it's possible in very small pieces)—these are challenges that should inspire a huge "so what?"

Those who design great treks across vast wastes would laugh at a challenge of standing on one foot for as long as you can or hopping around Manhattan with a paper bag over your head—yet each has exactly the same benefit: it tests the limit of human endurance. We should see these "great feats" for what they are: a way of showing off that has no more value than standing on one foot. With that viewpoint in mind, McKibben's concerns are irrelevant. But even if you do support the view that marathons and the like have benefit for pushing the limits, then it's just a matter of setting different limits. However far enhanced our bodies may be, there will still be limits to our capabilities. Even those who think they can make human bodies last forever (see page 36) don't think they can give them infinite power or invulnerability to all challenges.

This is not to say that we necessarily want to give parents the chance to make every possible choice in the genetic makeup of their child. Pushy parents don't generally make attractive viewing, and this would seem the ultimate example of the stage mom, not only pushing her children into a particular career but also designing them to be best suited for it. And just as there is doubt about the psychological effects of being a clone of another person—having to live with the knowledge of having been produced for a reason—so it might not be ideal for a child to grow up knowing that his or her parents always wanted a little runner or a little weightlifter.

However, the real world has an irritating habit of turning out differently from our intent. Human beings are by no means the pure product of their genes. That's why a clone will never be a "mini me"—cloned animals often look totally different from the animal they are cloned from and behave differently, too. While we could engineer a child to be extra strong, it wouldn't stop him or her from growing up wanting to be a ballet dancer or a truck driver. If we do ever manage to engineer in physical capabilities genetically, we aren't turning people into a particular profession or expertise; we are adding more tools to their resources.

Take the argument against giving people extra capability than they have "naturally" at face value. By the same argument, shouldn't we avoid educating our children as well? After all, think of all the problems that result from education. Some children will get a better education than others—surely that's unfair? And an educated child will be able to do things a "normal" uneducated child couldn't do. By teaching them math, think how we are removing from them the challenge and effort of working out their sums on their fingers. It's unnatural.

Of course this parallel sounds stupid. No one would suggest we ought to stick with the "natural" condition of not having an education and not being given tools with which we can achieve much more than we could if only left with what we are born with. Equally we shouldn't deprive anyone of the many extensions of human capability that are described in this book. Yet this is exactly the argument that is put up against genetic enhancement.

This doesn't mean we should rush in and try to enhance children genetically today. We just don't know enough. To make the kind of change that gives extra strength (for example) it's necessary to undertake germ line modification, altering the initial cell of the embryo, so that every cell in the body has the new genes. This means that these same modified genes will pass into the human gene pool. If there's a side effect—for example, if two generations after the change all people with this modification become infertile or start to die at age twenty—then we've injected this trait into the gene pool and it may spread (though like most genetic defects, it will tend to restrict itself unless the result isn't obvious until after childbearing age).

Any modifications would have to be intensively tested, and we would need to be happy that we can make genetic modification without causing collateral damage. Our knowledge certainly isn't up to that level at the moment. But assuming we did achieve this level, then the argument that we shouldn't enhance human beings because we would be losing some of the challenges we face is not an acceptable one. It goes against the whole nature of being human. We do enhance ourselves. Otherwise we would still be boring old Human 1.0, crouching on the savannah without a single tool or adaptation to help us.

Despite the predictions of some futurologists, it is never going to be the case that you could decide to have a daughter who grew up to be five foot four, with a particular pretty face that you picked out of a catalog. Genes provide the cookery recipe for producing a human being, but however good your cake recipe, it isn't the same thing as baking a great cake. You have to do the mixing right and get the temperature right and put the frosting on evenly and so on. It's the same with the human "recipe." Environment has a huge impact on our appearance and development. We would rightly be dubious about picking out a designer child this way. But I, for one, wish my parents had been able to arrange that I would have had less chance of getting fat and less chance of dying of heart disease than the genetic throw of the dice has allowed me.

Whatever is achieved through genetic enhancement, there are still going to be limits on our physical capabilities that come out of the gravitational pull of the Earth and the simple physics of scale. When we think of extra strength we tend to think of greater size—but as you make a creature larger, you also make it more vulnerable in some ways. Fiction has given us giant creepy crawlies to terrify us, from the massive ants of the 1950s movie *Them!* to the huge spiders in The Lord of the Rings and Harry Potter movies, but in reality they could never exist.

Imagine making a spider 100 times bigger than normal. When we say "100 times bigger" we really mean 100 times as tall or 100 times as wide. Its legs would be 100 times as long. But if we looked at the area of the spider's legs, that would be 100×100—10,000 times bigger. And what about the weight? Weight depends on volume, so that would not be 100 or even 10,000 times, but $100 \times 100 \times 100$—1 million times bigger. So you've a million times as much weight being supported by legs that are only 10,000 times larger in cross-sectional area. Result? The spider would collapse under its own weight.

It would also run out of breath, because it breathes through its skin—so it has 10,000 times more intake of oxygen to support 1 million times as much body. As animals get bigger, the legs have to get thicker much more quickly to be able to support the weight. We couldn't scale up indefinitely in size—and similarly there are limits to the changes that can be made to provide extra strength. This doesn't mean there isn't a lot we can do. We can, for instance, reduce the impact of physical tiredness through modification of our internal mechanisms. But we can't do anything about the basic physics of scaling, the strength of materials, and the work required to counter the pull of gravity.

Because of the limits to the enhancement of the body itself, many developments that human beings have undergone over the last thousands of years involved going beyond the body rather than making a direct modification. In principle every piece of useful technology extends human capability, even if we never come into direct physical contact with that technology. Street lighting, for instance, extends our ability to see from the daylight hours into the night, even though the technology isn't attached to our eyes. Heating and air-conditioning transform our ability to survive (or be comfortable) outside a very limited temperature range without us having to add or shed layers of clothing.

Covering every aspect of technology, though, would take many volumes rather than a part of a chapter. Rather than plow through many examples in a summary fashion, I would like to concentrate on three of humanity's greatest technological achievements that are particularly interesting from the point of view of upgrading—the dog, artificial light, and flight.

Including the dog may seem more than a little bizarre. How can a dog be a piece of technology? It's a living creature. Yet a dog has two distinct differences from the wolf, the wild animal it was bred from, that make it a great example of our tendency to reach beyond natural human capability. First, dogs have functions. They don't just exist alongside human beings but carry on activity on behalf of humans. And second, dogs were the first example of animal genetic modification, creatures bred with a particular intent in mind.

The reason the dog is a particularly dramatic first is that it was the original autonomous piece of technology—something that runs on its own, rather than being powered by a human being. It extends human abilities, initially in the field of strength, without us even having to touch it. Rather than finding a way to amplify their own strength, the developers of the dog decided to make use of a package of external power.

A dog can run faster than a human being. It has a much more effective sense of smell. Its jaws are more powerful and its fangs larger and more dangerous than a human's comparatively weak teeth. If you consider a hunting and protection dog—the two initial roles of "man's best friend"—as an extension of the human being, it makes a formidable weapon with a reach that can extend far beyond that of a thrown spear and an ability to work when out of sight, and it provides a confusing second source of danger for any attacker.

Because of their pack loyalty, dogs rapidly became more than tools, developing a close and complex relationship with their owners. That the relationship is complex can be seen in the way attitudes to the dog have changed with time and in different cultures. Though practically every civilization has made use of dogs, there have been widely differing views of their nature. Dogs are often viewed in Eastern cultures as dirty scavengers, and a lot of our invective involving dogs, inspired by biblical language, still labels them dirty, lazy, greedy, and shameless.

This didn't stop dogs being used in profusion, and by the late Middle Ages a strong distinction was growing between the "noble" hounds owned and kept by the aristocracy as house dogs and allowed freedom of the home and working dogs, treated with as little care as any other animals in the period. The distinction between pet dogs and working dogs is maintained to some extent to this day, though it is no longer reflected in a separation of breeds, as practically every type of dog is kept as a pet.

Historically the different breeds were selected on the basis of traits that made them best suited for a particular role. Heavyset mastiffs as guard dogs and hunting dogs, intelligent, gentle retrievers to bring back fallen prey, wiry terriers to go down fox holes or to take on rats, sensitive hounds to follow scent—like any flexible piece of technology, the dog was developed into many different models to suit varying needs.

Some of those uses are still with us today. Although the majority of dogs are now pets, working dogs still extend human capability, some in ways that couldn't have been dreamed of when dogs were first bred. After hunting and protection, dogs came to be used to pull small carts and sleds, to turn spits in the fireplace, and to track down criminals. On the farm, the dog became indispensable as a patient assistant in rounding up sheep. The hunting dog breeds diversified—no longer were they just assistant killers but split into hounds, pointers, and retrievers.

Most remarkable of all is the role that dogs have fulfilled as helpers to the blind, the deaf, and the disabled. This is over and above the role the animals play as companions to many otherwise lonely people. There is some evidence of dogs being employed to help the blind early in history. One of the murals found when the Italian town of Herculaneum was excavated, long buried beneath the ashes of the volcano Mount Vesuvius when it erupted in A.D. 79, features a blind person being led by a dog, while a medieval wooden plaque also shows a blind man being assisted by a canine helper.

The concept was mentioned in passing in a couple of nineteenth-century books, but no one seems to have taken it seriously until World War I. The earliest organized attempt to train guide dogs was in Germany in 1916, intended to guide soldiers who had been blinded in battle. This idea spread to America in 1927, when an American woman working as a dog trainer in Switzerland, Mrs. Dorothy Eustis, found out about the German work and wrote an article that was picked up by the first American owner of a "seeing eye dog," Morris Frank, whose dog was named Buddy.

Since then, thousands of people have been able to recover an active life thanks to guide dogs. I recently watched a guide dog lead its owner from a train to the exit of Paddington Station in London. Despite the milling crowds, ticket barriers, a "wet floor" warning sign, and a whole host of hazards that seemed to have been put in the way deliberately to make the task of crossing the station difficult—the noise, the smells from Burger King and Krispy Kreme outlets, and the nearby presence of the huge, noisy trains—the blind owner was able to make his way at normal speed across the station and on his journey.

More recently, guide dogs have been joined by other types of "assistance dog." Hearing dogs alert their deaf owners to audible signals that a hearing person would pick up and respond to. It might be a ringing doorbell or the sound of a reversing vehicle nearby. Although a hearing dog doesn't need to provide the same precision of leading, it does have to make sophisticated distinctions in the melee of sounds that makes up modern life. The third class of assistance dog is a service dog, trained to help those with physical disabilities that make it difficult to be mobile or to manipulate objects. It is quite remarkable to see one of these dogs operating an ATM on behalf of its owner. Many assistance dogs are Labradors or golden retrievers, although a range of other breeds have been used.

Of course the production of this remarkable piece of technology didn't begin with the intention of creating such a flexible helper. The chances are it all started by accident. Although wolves don't deserve a lot of the bad press they get—they rarely attack human beings, for instance—they would have been irritating scavengers that early man had to make an effort to scare off to stop them from stealing the remains of hunted animals.

It's easy to imagine those first, tentative steps away from the wolf's role as enemy. Perhaps it was a cold winter and a wolf crept close to a fire to keep warm. Maybe while it was there some other predator attacked the camp—the wolf, ever the pack animal, jumped to the defense of the humans, fighting alongside them. It was rewarded with a gift of meat. Natural selection takes things forward from here. Over the years, wolf cubs that are more docile, more easily fitting with a human "pack," are the ones more likely to stay around and more likely to be fed and encouraged. Over tens or hundreds of years, the dog emerges.

Remember the experiment by Dimitri Belyaev that was mentioned in the first chapter. In just forty years he managed to turn silver foxes into something approximating dogs. It really doesn't have to take long. Perhaps one hundred years after that first tentative contact the early hunters were no longer dealing with wild wolves. The animals that lay around their camp had changed in manner and appearance. Their once-upright ears had drooped. Their coats were more varied in color. They accepted humans as part of their pack. The dog had been created.

This was genetic engineering, just as much as any GM crop, though the approach taken was more indirect. By selecting, consciously or unconsciously, for certain traits, humans have modified the nature of many animals and plants to better suit their requirements. This is particularly obvious in two plants, the cauliflower and maize. The cauliflower is a mutant cabbage. Its flower has been transformed into a hard, bumpy white structure—the part we now eat. With no functional flower, it can't breed without help. Similarly, maize has been selected over the years for bigger and bigger seed husks. It is now incapable of self-seeding and won't grow without human assistance.

Just as these plants are no longer viable in the wild, the dog is not a natural animal. It is as a much a human-made piece of technology as a table that started off as a "natural" piece of wood. Without doubt, the dog—along with the other animals and plants we have engineered over the millennia to better fit our requirements—is one of the most impressive ways we have gone beyond the basic human. Forget Stonehenge—it's a toy by comparison. Okay, it gives a handful of people some astronomical information and it's pretty—but it hasn't been in use for thousands of years. The dog is a piece of Stone Age technology, developed thirty-five thousand years *before* Stonehenge, and it's still going strong.

The wolves that eventually became dogs may have been attracted by the warmth of human fires, but the light of those fires would prove just as important to humans as the heat they provided—they were the first form of artificial light, the second of my key technologies. The Sun is the driving force behind life on Earth. Without sunlight we would freeze, we would have no atmosphere, and there would be no weather. Most important for the consideration of artificial light, without sunlight we would be unable to see—and sight has been fundamental to most of the superevolutionary development of humanity. In our natural state, when darkness falls we are vulnerable. We are unable to perform anything but the most basic tasks. Active life, and the ability to upgrade ourselves, is limited to the hours of daylight—these can be precious few hours in the winter months. Artificial light—and particularly the electric light—has transformed this.

Without artificial light, we would be unable to have any activity during the night hours. Much of modern life would be impossible without that illumination. Not just reading in the evening or watching TV or a movie or eating a meal or talking with friends but *any* activity beyond huddling for safety in the darkness. Without artificial light, materials could not be mined underground, most medical operations would be impossible, and half of our lives would be thrown away.

From the earliest use of fire, humanity had some degree of control over darkness, and all the candles and torches and oil lamps and gas lamps over the years have done little more than make the light of a flame more easy to control and more intense. The principle is still the same. Heat a substance enough and electrons in the atoms that make up that substance are stimulated to jump up to a higher energy level (they make a quantum leap). A little later the electrons drop back down and in doing so have to give off packets of energy equivalent to the change in level. These packets are photons of light—the result is that the burning object glows.

Candlelight and oil lamplight may be very romantic, but their flickering flames lack practicality as sources of illumination. They produce relatively little light, the flames waver in the wind, and before long the fuel burns away and the light dies. They just aren't good enough to enable us to fully extend many of our activities, from working in a factory to driving a car, into the darkness. Gas lighting took things further, producing light at the turn of the tap, but the principle was still the same inefficient approach as a candle or oil lamp—light from a flame. The only significant breakthrough in the technology in thousands of years was the introduction of the mantle. These fine meshes of metal or specially treated fabric were heated to white-hot by the flame, producing a brighter, more even light.

The electric light was to change all of this. Once Edison had developed the incandescent bulb in 1879 (after a little problem over patents with the scientist Joseph Swan, who arguably beat Edison to it), lighting was literally available at a flick of the switch.

The way that electricity is transmitted has changed from Edison's direct current, where one wire carries a constant voltage, to Tesla's alternating current that we use today, which swaps the power between two wires, carrying an ever-varying voltage. Edison, never one to give up a fight easily, tried to prove that AC was more dangerous than DC by killing animals with it—but AC is more cost effective, losing less power when transmitted than DC, and inevitably became the electrical mode of choice. However, the principle of the electric light was the same whatever the current. A filament (originally carbon, later metal) is heated until it glows brightly. It is only prevented from burning up altogether by the inert gas that surrounds it. Now, though, more and more artificial lights use a principle that was developed before these "incandescent" bulbs.

Fluorescent lighting, the technology used in modern low-energy bulbs, dates back to experiments by the German scientist Heinrich Geissler in the 1850s, over twenty years before Edison's triumphal introduction of his lightbulb. When an electric charge is passed across a low-pressure gas in a tube, the result is an eerie glow. Unfortunately, this glow is weak, which is why Geissler's tubes never caught on, but in a fluorescent tube the brightness is magnified by using a coating that gives off visible light when hit by ultraviolet, converting the mostly invisible output of these discharge tubes into a strong, clear light.

Artificial light gives us a powerful strength we otherwise lack. Naturally, most animals tend to be either daytime oriented or nocturnal. For early man, night was a time of mystery and fear. Artificial light has given us a way to turn that night (locally at least) back into day, to remake time to order. Our lights make it possible to explore where we otherwise would not dare to venture. It means we can cross continents in the darkness—and use light-based media such as movies, TV, and computers. Artificial light is special because, despite being intangible, it expands the human capability across a much wider scope.

Last, but by no means least, comes flight. Many millions of years were involved in the evolution of wings by birds, insects, and a few small mammals. Even after all that time, a bird can take weeks over its migration journey. Thanks to only a couple of thousand years of technological development, without the slightest biological modification, I can fly that same distance in a matter of hours. Natural evolution takes so long because it is driven by random steps. It is blind—there is no sense of direction. Our unnatural evolution operates on a different timescale because it is directed by purpose. We might not know how to get there, but we know what we are aiming for.

The urge to fly goes back a long way. While the technology didn't exist, there's no doubt that for millennia people dreamed of being able to soar into the air like a bird, to travel from place to place avoiding the barriers of rivers and lakes and mountains. In the earliest legends involving taking to the air, the natural route of harnessing creatures that can already fly was common. A Persian king by the name of Kai Kawus, according to legend, hitched up eagles to a chariot and soared away around thirty-five hundred years ago.

Other legends left the flying to manpower. Perhaps the best known ancient tale of flying was the Greek myth of Daedalus and Icarus. Daedalus was an inventor, said to have designed the labyrinth for King Minos of Crete. This complex defensive maze contained the half-man, half-bull monstrosity the Minotaur. Architects of secret structures of the time were sometimes killed to destroy their knowledge. It might have been better for Minos if Daedalus had suffered that fate. The inventor gave the secret of getting through the labyrinth to the king's daughter, Ariadne, who passed it on to her lover, Theseus. After Theseus managed to kill the Minotaur and escape, Daedalus was imprisoned along with his son Icarus. Daedalus built wings of wax and feathers so the two of them could fly away to safety, but Icarus was too bold, reveling in the freedom of flight as he soared higher and higher. Forgetting his purpose, he trespassed onto the territory of the Sun. The heat of the Sun's fierce rays melted the wax on his wings, leaving Icarus to plunge to his death in the sea. This story was designed to be a pointed parable, demonstrating the dangers of knowledge and of wanting too much for the self. Like many myths, it was never intended as anything approximating the truth; it was a story to educate—even so, many still believed it was only a matter of building the right wings and men would fly like birds.

There was also a story going around in ancient Greece that Alexander the Great followed in the footsteps of Kai Kawus but went one better than the eagles, fastening a flock of griffons to a wicker basket, in which he was hauled into the sky. Given that griffons (also known as griffins and gryphons) never existed, being a mythical cross between a lion and an eagle (a combination that would find it impossibly difficult to lift their own body weight, let alone Alexander's), this is not the most likely of stories. In fact, legends like this, and probably that of Kai Kawus, too, are more symbolic of the general or king's farseeing wisdom than ability to fly.

An interesting parallel emerges when the medieval proto-scientist Roger Bacon talked about using lenses to see at a distance. He ascribed an early use of this technology to Julius Caesar: "So, it is thought, Julius Caesar spied into Gaul from the seashore and by optical devices learned the position and arrangement of the camps and towns of Brittany." This was a modernization of the same kind of legend. Great war heroes had to be farseeing. In the early days, this was attributed to being carried up into the sky by animals—Bacon had Julius Caesar using science. In reality, Caesar's success was all down to luck and cunning, but we like to find explanations for exceptional performance. We talk of a great general when in truth we mean a lucky one.

In the Chinese Taoist tradition, the aim was to become a *hsien,* a person who is immortal and can fly in the heavens, riding the wind. In the tales of the Middle East we hear of flying carpets, able to carry the owner wherever he or she wants to go. On a more scientific level, Roger Bacon, when listing the wonders that science and human ingenuity can produce, includes this rather startling statement for the thirteenth century: "It is possible that a device for flying shall be made such that a man sitting in the middle of it and turning a crank shall cause artificial wings to beat the air after the manner of a bird's flight."

This is quite different from the attempts of the monk Eilmer (often called Oliver due to an early mistake in copying his name) of Malmesbury, who in 1010 undertook an experiment that Bacon was probably aware of. Eilmer tried to follow Daedalus and made himself a huge pair of feathered wings. Such was Eilmer's faith in his technology, he threw himself off the top of the abbey church tower. There must have been some gliding capability in the wings, as he survived, though he did break both his legs.

While it's true that the device Bacon describes sounds more like a real attempt at a flying machine than Eilmer's wings, even though it could never work in practice, he does give us a word of caution: "These devices [the flying machine was one of many wondrous engines he described] have been made in antiquity and in our own time, and they are certain. I am acquainted with them explicitly, except with the instrument for flying, which I have not seen. And I know no one who has seen it. But I know a wise man who has thought out the artifice."

Many men, wise or otherwise, would try to bring the idea of human flight to life over the following centuries (women, on the whole, seem to have had more sense). Two who are notable for their timing were Senecio of Nuremberg and Giovan Battista Danti da Perugia, who both followed Eilmer of Malmesbury in their attempt at soaring. Neither was a great success. As a character in *Monty Python's Flying Circus* comments, when attempting to get sheep to fly by pushing them out of trees, "They don't so much fly as plummet."

The reason Senecio's and Danti's attempts in 1496 and 1503, respectively, were significantly timed is that they occurred during the lifetime of Leonardo da Vinci. There is no evidence that Leonardo ever built a flying machine, but he put a lot of effort into designing them. Leonardo's first talent was art. There was no reason that a fourteen-year-old from a rich family in Florence (the family moved from Vinci while he was still young) should take on an apprenticeship unless he already had a special talent, but this Leonardo did in 1466, under Andrea del Verrochio. For twelve years Leonardo was to work in the studio of this master painter and sculptor before launching his career with a first commission to paint an altarpiece for the

Florence town hall. Like many of his works, this was never finished.

Leonardo didn't stay in Florence long once he became independent. Instead he wrote what must be one of the most deceptively self-flattering letters ever to the Duke of Milan, claiming to be a military and nautical engineer of unrivaled excellence. Leonardo, with no practical experience under his belt, got the job on the basis of this wonderful piece of bluffing but must have already had an interest in things scientific. As the duke's principal engineer Leonardo showed amazing flexibility in the speed with which he picked up technical knowledge.

It was with his engineering hat on that Leonardo devised different ways to handle manned flight. He sketched an experiment to test how effective the lift of a wing would be (like many early attempts at producing wings, this looked like something off the set of a Batman movie, with birdlike struts and scalloped wing edges). He drew several devices that resembled the one Roger Bacon mentions, down to the cranks to make the wings beat. And, famously, Leonardo devised a kind of early helicopter, cutting through the air like a helical screw. Altogether there are an amazing 150 different flying devices scattered among Leonardo's notebooks—this was certainly an important endeavor for him.

The failing birdmen would keep at it all the way up to the twentieth century. Even today, in the seaside town of Bognor, England, the International Birdman Festival annually features a procession of individuals from around the world who are prepared to jump off a pier into the sea in an attempt to win a fifty-thousand-dollar prize for the farthest distance flown beyond 100 meters (the record is currently 89.2 meters), supported by whatever crazy outfit they hope will keep them in the air long enough under their own power. But as often happens in science and technology, the first effective manned flight came from a touch of lateral thinking. There were no wings involved at all.

The problem with trying to fly like a bird is the relationship between weight and thrust. Birds' wings can lift them off the ground because they have (relatively) very strong wing muscles and, crucially, are very light for their size. It's partly a problem of scale. Smaller things can get away with what would be amazing feats of strength on our scale. Trying to emulate a bird in normal flight was a nonstarter, and though there were some rumors of successful attempts at gliding, the one mode of bird flight that *is* possible for a human with big enough fixed wings attached, the first real success in the air came from taking the creative step of disposing of wings altogether. If the Montgolfier brothers had any inspiration from nature, it wasn't from birds or bats but from sea creatures, some of which can rise or sink by changing their density relative to the water using a special bladder. The French pair came up with an equivalent aerial device, the balloon.

Joseph Michel and Jacques Étienne Montgolfier owned a paper mill near Lyon in France. They weren't the first to notice the way that paper was lifted by the hot air from a fire (or even a candle), but they did take the big step of going from playing with an open paper bag to producing larger and larger balloons, capable of lifting increasingly large payloads. As in the space program, the first passengers in a balloon were not people but animals. On June 4, 1783, a thirty-two-foot-wide balloon was launched from the royal palace of Versailles carrying two birds (a goose and a cockerel) along with a more unlikely aeronaut in the form of a sheep. The animals were brought back to land safely after flying for around two miles.

Five months later, on November 21, 1783, men followed in the footsteps of the sheep and birds—a scientist and a soldier, Jean François Pilâtre de Rozier and François d'Arlandes, became the first human beings to enjoy true flight in a seventy-two-foot-high balloon. The king of France, Louis XVI, was so impressed with this feat of private enterprise that he ordered his resident scientist, Jacques Charles, to make a manned balloon for him. Charles had not seen the Montgolfiers' balloon himself, so was not aware that it used hot air to provide lift. But he did know about the recently discovered element hydrogen, which was much lighter than air. When Charles built a balloon for the king, which took off from the Tuileries gardens in Paris less than a month after the Montgolfiers' first manned flight, on December 1, 1783, traveling over twenty-six miles, it used hydrogen for lift.

Balloons would be increasingly popular through the next century and into the twentieth century, culminating in the massive airships such as the 804-foot-long *Hindenburg*, used for transatlantic flights. Technically, the airships were dirigibles—balloons built on a rigid structure rather than inflated by the gas that kept them aloft. Like Charles's balloon, the great airships were still filled with hydrogen, a very flammable gas, leading to the disaster that wrecked the *Hindenburg*, killing thirty-four people and destroying the future of airships as commercial vehicles. They have had a small resurgence now that they can be filled with helium, a gas that is still very much lighter than air but doesn't burn, but they have never recovered their importance in aviation.

Through the nineteenth century it became increasingly obvious that it wasn't going to be possible to use an ornithopter—a flying machine that worked by flapping its wings like a bird. In the early 1800s, English inventor Sir George Cayley suggested the only way forward was to combine rigid wings with some means of propulsion like a propeller (already in use on ships), while around 1840 William Henson devised such a craft using a steam-powered propeller (though it was never built). Surprisingly, although there is anecdotal evidence of earlier examples, it wasn't until 1894 that we know for certain anyone managed to fly a fixed-wing glider. This was the German inventor Otto Lilienthal, who had been fascinated by the idea of flying since he was a boy. By 1894 he was airborne, originally using bird-shaped wings, which later became more stylized with the addition of a second wing and tail. Lilienthal was certainly successful but was killed in flight in 1896 when his glider flipped over, fifty feet above the ground.

In 1903 a new era began with the advent of powered, heavier-than-air flight, and the honors went to the United States. The two brothers Wilbur and Orville Wright ran a bicycle repair shop, but even though the construction of an aircraft was effectively a hobby, they took it very seriously. Rather than rely on birdlike wing designs, they built an early version of a wind tunnel and used it to test out different wing shapes to find the airfoil that was most effective. Their new wing shape was then tested out as a glider before they went on to make the powered version, using a tiny four-cylinder internal combustion engine they had designed themselves.

The Wrights' first attempt with their Flyer on December 14, 1903, at Kitty Hawk failed. They tried again on December 17, this time launching from a rail rather than using helicopter-like skids. With Orville on board, the Flyer covered 120 feet during its twelve seconds in the air. Travel would never be quite the same again. In just over a hundred years, we have gone from the fragile, slow, string-and-sealing-wax early flying machines to airplanes that carry hundreds of passengers at near the speed of sound (twice the speed of sound until Concorde was retired), miles above the ground. Flying has become routine, even something that causes a touch of guilt because of the amount of global warming it produces—but we should never forget what a remarkable achievement the Wright brothers made back there at Kitty Hawk.

Of all the technologies, flying is the most dreamlike, propelling us into a fantasy world. In evolutionary terms, the move to flight has been hugely costly for the animals that have adopted it. To be able to take to the air, they have undergone vast changes compared to the nonflying creatures they evolved from. We have stayed biologically the same but have gained the amazing ability to get from A to B at over one hundred times our normal speed, passing over any obstacle as if it weren't there. For all the mundane nature of flying today, despite any doubts about its impact on climate change, it is still a magnificent achievement and a great step forward in the upgrading of human capability.

Back in history when the idea of people flying seemed more suited to myth than reality, it was more reasonable to think of items to wear that would enhance our natural abilities, so we come full circle from external devices like aircraft and dogs that take us beyond our limits to worn technology that more directly enhances our inbuilt physical capabilities. The traditional fictional way to exceed the normal span of our ability to walk and run was pulling on seven-league boots. (This was echoed in twentieth-century popular culture when Jerry Siegel and Joseph Schuster dreamed up Superman, who originally could leap tall buildings at a bound rather than being able to fly.)

The concept of seven-league boots was simple, even if the mechanics of using them was never clearly explained. A normal stride is around a yard. With these magical boots in place, somehow every step you took would cover seven leagues. A league is an imprecise measure that comes in somewhere around three miles, so that gives you twenty-one miles for each pace—not bad going. If you took a sedate stroll at one step a second, this would amount to traveling at around 75,000 miles per hour. If it weren't for the sea, you could get all the way around the globe in less than twenty minutes. Of course, you would be traveling much faster than the speed of sound (around 700 mph) so would leave a sonic boom in your wake. (This assumes that you really had a twenty-one-mile step at a normal pace. I've never been clear how the wearer of seven-league boots progresses: perhaps the idea was more like floating over that twenty-one-mile gap, so you proceed in huge hops, rather than even paces.)

Seven-league boots may be fantasy, but there is now a way to extend your stride using PoweriZers, spring-loaded stilts that strap onto the wearer's legs. PoweriZers (and competing products) allow the user to jump up to six feet in the air and to run with nine-foot strides—not exactly seven leagues but far exceeding the normal capability. They take some getting used to, but expert wearers can perform gymnastic feats that are impossible for the unaided body.

A company called Applied Motion has gone one step further (almost literally), building sprung stilts into a frame that straps onto the body. The original version of the SpringWalker, a product that the manufacturer refers to as a body amplifier, is powered by human energy, using the legs and arms, but the aim is to have externally powered versions available relatively soon. Like

PoweriZers, the SpringWalker gives the wearer the ability to move at an unusually fast rate, which in the powered version will also be effortless—but without the normal problems of wheeled vehicles when faced with obstacles.

In essence, the SpringWalker is a simple mechanical exoskeleton. There is nothing new about the concept of a creature with an exoskeleton. It's just an animal with a hard structure on the outside, rather than the inside where creatures like us keep our skeletons. There are many more animals on this planet with exoskeletons, from ants to lobsters, than there are with a backbone like ours. But if DARPA, the U.S. Defense Advanced Research Projects Agency, founded in 1958 as a response to the Soviet launch of the *Sputnik* satellite, has its way, in the future we will go far beyond the SpringWalker, not only extending our stride but also expanding the whole physical capability of our body and doing so with serious horsepower, not just the puny leverage of a spring. Inevitably with military applications in mind, DARPA has plowed millions of dollars into the development of military battle suits in its Exoskeletons for Human Performance Augmentation program.

DARPA's vision is to give a human being a metal external skeleton that can be powered to provide immense strength and speed. This isn't the same as a cyborg, a machine/human hybrid (see page 206), but rather a powered suit, worn by the human who controls it. Exoskeltons have been around a good while in fiction, from the fighting machines of Zion in the Matrix movies to the complex walking battle engines of computer games such as *Heavy Gear* and *MechWarrior,* but unlike humanoid robots, where the fictional version is way beyond our current reach, much of the capability of an exoskeleton can be achieved with relatively simple mechanics (and a touch of hydraulics and gyro-scopes, plus computer power to provide control and maintain balance, thrown in).

In one small way, the movie industry has its own exoskeleton, which has transformed handheld camera work—the Steadicam. Until this device came along in the 1970s, handheld cameras were shaky and amateurish, but a camera on a tripod or a wheeled dolly hasn't got the flexibility of movement available to a human operator. The Steadicam works by putting the camera on one end of a boom, with a counterweight (usually including the camera's battery) on the other end. The boom is pivoted in between the two weights on a gimbal, which connects the camera boom to a harness worn by the operator. The result is a camera that floats in space, almost independent of the movements of the cameraman. All his hand does is steer the mechanism—the cam-era is held in the air by this simple exoskeleton.

In the last few years, DARPA's plowing of money into exo-skeleton research and development has started to pay off in the form of working prototypes. The aim, as much as possible, is to mimic the motions and capabilities of the human body, particu-larly in carrying and moving, but with strength or speed that far exceeds human capabilities. The most advanced of the current models use a hydraulic system, employing a fluid link to amplify any inputs.

A simple example of a hydraulic system is the hydraulic press, which like many other water-powered devices was first used in the nineteenth century. The system works by enclosing a body of fluid in a cylinder with two pistons, one with a small surface area, the other much larger. Pushing the small piston into the fluid re-sults in pressure on the large piston, which moves out as the smaller piston moves in. Because of the difference in volumes of fluid displaced, the large piston moves less far but feels a proportionally larger pressure. So a small force on the small piston is turned into a large force (working over a much smaller distance) on the large piston. It's the fluid equivalent of gearing.

Such a system can be used to turn even manual effort into greatly increased force, but in the prototype exoskeletons the source of the power is more likely to be either electrical or an internal combustion engine. The electrical option is the most desirable and flexible—there is something endearingly makeshift about having to start a gasoline engine on an exoskeleton as if it were a lawn mower—but the limitations in current battery technology mean that the electrical exoskeleton, just like the electric car, is limited in its capabilities.

An exoskeleton that runs out of power halfway across a battlefield would not be popular. DARPA's aim is to have a device that will "assist our soldiers in carrying [their load of armor, weaponry, and supplies] by developing a fully integrated exoskeleton system that will increase the speed, strength and endurance of load-burdened soldiers in combat environments." The idea is to have a heavily armored exoskeleton that is designed to have mission-oriented packages attached to provide the weaponry and capabilities required in different environments. When the exoskeleton is working, the operator will be able to move naturally, unencumbered and without additional fatigue, while the machinery carries the payload. Just how much progress has been made isn't clear, but DARPA says that the project is in its "final phase."

One of the earliest examples of a practical exoskeleton emerged from the University of Calfornia at Berkeley. In 2004 their Robotics and Human Engineering Laboratory demonstrated BLEEX, the Berkeley Lower Extremity Exoskeleton, a pair of powered metal leg braces connected to a backpack that enables the wearer to carry heavy loads over long distances on foot. The initial version made the one-hundred-pound device, plus a seventy-pound load, feel to the wearer as if he or she were carrying no more than five pounds in weight.

Although military exoskeletons, with their sci-fi connotations, tend to grab the headlines, if only because that's where the most money is, there are also real possibilities of exoskeletons being used to help the disabled where limb or nervous system damage makes it impossible for individuals to use their own muscles. In 2006 Panasonic demonstrated a compressed-air-driven "power jacket" that amplifies the movements of arm muscles to help patients recover from partial paralysis. Unlike military exoskeletons, the system itself is very light at only four pounds, though it does require a connection to an external compressed-air device and is more a treatment than an augmentation for everyday living—yet it demonstrates how this kind of technology can also have medical applications.

If Professor Yoshiyuki Sankai of the University of Tsukuba in Japan has anything to do with it, exoskeletons will be commonplace sooner than we think. He has joined forces with Daiwa House Industry Company with the aim of making around four hundred powered suits in 2008, which will be leased out to medical companies to help elderly and disabled people deal with everyday tasks. Rather than amplify existing muscle movement, these suits are designed to "read signals from the brain," so should be suitable for those with difficulty moving their limbs, though it's a steep challenge to make them work, given all the difficulties of mind–machine interface (see page 231), and Professor Sankai could find that his promised technology takes longer than expected to develop.

One of the problems with using hydraulics or compressed air to enhance the human musculature is that the mechanisms can be bulky and take up a lot of room. The Panasonic jacket, for instance, despite being lightweight, has various tubes sticking out of it and requires connection to a relatively large pump system. Researchers at the University of Texas at Dallas in Richardson have been taking a different approach, aiming to provide an artificial equivalent of the muscles in the human body.

Two early versions of an artificial muscle up to one hundred times stronger than the natural version have already been built. One is powered by alcohol. The "muscle" is a length of wire with a special catalytic coating that causes the fuel to burn flamelessly, heating the wire, which contracts, producing a pulling force. The second uses carbon nanotubes—extremely strong and thin tubes made from lattices of carbon atoms, which change shape when an electrical charge is passed through them. The carbon nanotube "muscle" works without a battery, as a catalyst is used to produce a charge from a reaction when oxygen reacts with the fuel. What is particularly clever about these artificial muscles is that they are both the actuator that makes the movement happen and the power source—just like a physical muscle—so are much more compact than conventional mechanical devices.

Upgrades have vastly enhanced our physical strength and our weaponry, and there is no sign of that stopping. There was one outstanding weapon, though, that came with the initial package of a human being. The deadliest weapon of all—but still one that was capable of enhancement. Not only did our enlarged brains enable *Homo sapiens* to begin improvements to the species; those brains would eventually turn their evolutionary capabilities on themselves.

5.
The Deadliest Weapon

Some recent philosophers seem to have given their moral approval to these deplorable verdicts that affirm that the intelligence of an individual is a fixed quantity, a quantity that cannot be augmented. We must protest and react against this brutal pessimism.

—Alfred Binet, *Les idées modernes sur les enfants*

If you've ever said, "I just can't get started without a coffee," you aren't alone. Around the world, millions of people are using the drug caffeine right now to make changes in their nervous systems. Their aim is to give a little boost to the deadliest weapon on the planet—the human brain. By comparison with many predators, a human being is physically weak, but lurking behind all the self-development circumventing evolution that is covered in this book, pulling the strings, is our intelligence. Brains, properly used, have become the ultimate tool for survival of the species. So it's not surprising that from early days we have looked for ways to extend the brain's capabilities.

I'll go back to early times in a moment, but let's revisit that coffee first. The caffeine in coffee, in caffeinated soft drinks, and (to a lesser extent) in tea is a drug that enhances the performance of the brain. It should come as no surprise that coffee keeps us awake, but it also has a small effect in increasing the ability to concentrate and tightens up reaction time a touch. Stimulus from caffeine seems to go back a long way, with the use of tea as a stimulant going back several thousand years in China, while coffee gradually filtered into the West from Africa in the sixteenth century.

Caffeine has several mechanisms that act on the body, but the key component for our thinking power is the way that it locks onto the receptors in the brain that usually handle a chemical called adenosine. The interference of caffeine reduces adenosine activity, which results in an increase in the activity of another natural brain chemical, dopamine. This is a neurotransmitter, one of the molecules used to help carry signals from neurons to other cells. The result is the stimulation that we all know. It's quite scary that something as simple as drinking a latte or sipping a cola can result in changes in such a fundamental operation of the brain.

Stimulants like caffeine may have been used as long as humans have been aware of the effects of chewing a certain leaf or brewing a particular drink, but some of the external enhancements to the brain's basic functioning have been around much longer. One of the first of these, a strange hybrid of external and internal influences, was language.

Language is so much a part of life that it seems strange at first to regard it as something more than a basic part of being a person. It might seem that language is something that is as much a biological part of us as is, for example, being able to pick things up or touch things. Yet in reality, our complex use of language is at least in part a bolt-on, something extra added by our extraordinary brains.

This isn't saying that communication is unique to humans or that it was missing from the "design" when Human 1.0 first evolved. Everything from bees to gorillas is hugely dependent on communication with other members of its species. Nor is there anything special about communicating by using sound, as we do by default when using language. It's probably the most common vehicle for communication among animals. But there is a fundamental difference between normal animal communication and language that reflects our upgrade from our basic biological form.

Communication below the level of language is immediate. It indicates something in the here and now. It can arouse emotions and trigger actions, but it can't foster ideas, plan for the future, engage in "what-ifs," convey knowledge, or spark invention. Language is required to enable practically all of the other advances we have made away from our biological prisons. Language is the ultimate catalyst of change and as such is a hugely significant part of going beyond the basic human capability.

It's impossible to say with any certainty when language began. More so than practically any other example of our extension of humanity, it leaves no trace in its earliest direct forms, only being directly preserved once it is conveyed in solid form as writing. It could have been anything from 2 million years ago, before the emergence of *Homo sapiens*, to forty thousand years ago that anything like true language developed. If the former, language could be seen as a part of the original "human specification" rather than something we've added on. If that's the case, then like our big brains, it's an essential prerequisite to being able to upgrade ourselves; if it's not, then it has to be about the most important of our upgrades—either way, it is fundamental to the process of going beyond biological evolution.

It certainly seems to be true that our brains have inherent capabilities for language—language "modules," as cognitive scientist Steven Pinker has called them—yet this is a more abstract concept than many of our other skills. What Pinker and others like American linguist Noam Chomsky are saying is not that language itself is inherently built-in. The sheer variation of languages around the world (as opposed to the universality of some preprogrammed facial communication expressions like fear or happiness) makes it obvious that we aren't hardwired for the specific details of a particular tongue. But all languages have some basic features in common, particularly the need for grammar and syntactical structure. It is just possible that this is a result of all languages having some form of common ancestor, but it's more likely that this is because there is a good fit between these aspects of language and the modules in the brain that handle communication.

I ought to qualify this term "module." There is no suggestion from Pinker et al that there is a specific, individual part of the brain that is the "language module" in the same way that the hippocampus seems to be the "storing memories module." Rather there are separate parts of the brain that come together to operate as a functional module to deal with language, providing a universal language handler that is then brought into use to deal with local specifics required to speak in English or Swahili or Japanese—or whatever language is required.

In the past, language has been as much of a barrier to human enhancement as an advantage because of the vast spread of different languages around the world. At one point this variety was even greater. Imagine people in every county in the United States speaking a different language or a variant of English so impenetrable that the people from the next county could only understand a few words. While this kind of diversity is fascinating and important from a cultural standpoint, it gets in the way of our ability to share information, without which there isn't much chance of upgrading. Unless an idea can be spread, it won't make a change to humanity as a whole, and that spreading generally requires language. There have been two significant contributory factors that have helped overcome this debilitating diversity.

The first unifying principle was Latin. As the language of the Roman Empire it was well spread through the Western world and was to have a much more lasting impact than the Romans themselves. Latin was the universal language of education in the West all the way through to Newton's time. The use of Latin meant that scholars could move from university to university in different countries and join in debates freely or read manuscripts written by other scholars without language forming the horrendous barrier that it does to this day in the political arena.

The other, and even more effective, language to break through as a way to provide worldwide communication—the one that makes possible the rate of change that we see now—is English. This had three things in its favor. The first was the British Empire and its former colonies, reaching across parts of the world that the Romans never even envisaged. The second was American commerce and media that has spread the use of English to make it truly global. Third, clinching the success of English, is the Internet.

In some cases where a single language is essential—for example, for use by flight controllers communicating with aircraft—English has been the standard for many years. In academia and to some extent business it has taken on the same role. It is a real shame that politics hasn't caught up, requiring all the expense and opportunity for misunderstanding generated by the language-to-language translation used in the United Nations and the European Union. Perhaps the answer might be to rename English something with no association to the United States and the United Kingdom. There is little doubt that language has been a major spur in our ability to upgrade and the availability of a near-universal language has been responsible for the acceleration of that upgrading in recent years.

There is one other universal language that has been incredibly important in our ability to advance—mathematics. Like language in the normal sense, math is something that we have some innate abilities for. Even some animals have limited mathematical abilities. We know, for example, that dogs can count up to small numbers, as they express surprise when tricked into thinking there will be (say) two items in a bowl and there are actually three. But counting physical objects isn't really the point at which math launches off as a powerful part of the foundation of our more-than-human abilities. It's when the idea of numbers and the manipulation of those numbers become abstracted from counting physical things that the power of mathematics really comes into its own.

Even as simple an idea as a negative number (without getting into more abstruse mathematical concepts like the imaginary numbers that are essential for much of physics) is quite an abstraction. I can show you one apple or two apples. I can kind of show you zero apples (though this is already a little vague as a concept). But I can't show you minus two apples—just the result of an operation that involves a virtual minus two apples, such as adding minus two apples to three apples and leaving one behind.

Language (and math) was an absolutely fundamental component in the human ability to drag ourselves up by our own bootstraps and go beyond the biological basics of evolution, but its basic form, speech (along with visual analogs like sign language), has one problem—speech is ephemeral. Once a piece of information is spoken, without any way to preserve it, it only continues to exist as long as the memories of those present preserve it or until it has been respoken, often in an embellished form. In a sense, a degree of natural selection comes to play here. We can remember elements of a stirring speech that we heard weeks ago, but we forget what someone said at the water cooler this morning before the day is out. Even so, left to its own a memory of a spoken piece of information is very fragile.

It is arguable that it is from the need to preserve the spoken word that our delight for storytelling emerged. It is much easier to remember a story than a collection of isolated facts. The oral storytelling tradition may now be largely regarded as an art form, but initially this wasn't its fundamental purpose. Storytelling was the first information technology. It was a way to preserve information and pass it on that both was memorable and enticed the listener into carrying on listening. Storytelling is still a great way to get information into people's brains—but it isn't ideal as a way of preserving absolute facts.

Even the best storyteller can occasionally forget things and add in details, so that a story will gradually modify over time as it is passed from person to person. The creative human brain is good at finding solutions around us to overcoming a problem like recalling the details of a story. If a narrative is too difficult to remember in total, perhaps some of it could be recalled by using a picture or just a mark in the sand or on a wall to trigger a memory. This way the long journey from memory to the widely available written word began.

In terms of its impact, its sheer transformation of human limitations, writing is phenomenal. In physical terms, writing can be as simple as a few marks in the sand or a blob on a piece of paper, but in conceptual terms, writing frees up communication in time and space, destroying the shackles of here and now.

Most animals and even some plants communicate at some level—but usually that communication is immediate and then it is gone forever. Scent-based animal communication lasts a little longer. The sort of spraying cats do to mark their territory might last a week or two, but then disappears and has to be restarted from scratch. Spatially, even this form of animal communication is limited to one place—a good thing, as it happens, for the cat, which wants to mark a single location, but restrictive in terms of communication.

Writing takes away the limits of space and time. I can take a book off the shelf and read words written thousands of miles away and hundreds or even thousands of years ago. There are probably more communications on my bookshelves from dead people than from the living, and certainly very few of the writers live near to me (the only one I can think of is fantasy/satire writer Terry Pratchett). On my computer I can read an e-mail written in the middle of the night my time that originated on the other side of the world. When you read these words it will be months or years after the moment (11:36 A.M. GMT on Thursday, May 24, 2007) when I write them. The chances are that you are hundreds or even thousands of miles from my desk. It doesn't matter. Writing takes care of time and space.

Of course we now have many other ways to communicate that are more instant than writing, but often these instant communications only manage to share the spatial aspect of writing's power—they often don't overcome time. Because they are written down, these words will still be here in ten years' time, maybe even a hundred years' or a thousand. The cold-selling phone call I just received from a stockbroker in New York was instantly consigned to the bin of time—the communication is as dead as a dog's bark (thankfully, in this particular instance).

Writing is a crucial accelerant in the development of human enhancement. We would still have gone beyond biological evolution without the written word, but the rate at which things have taken off in the last two and a half thousand years is largely due to writing. Without writing there would be no science, only myth. With no way to build practically on the experience of previous years, we would always be reinventing the wheel. Computer technology is often seen as something of an enemy of writing—why read a book when you can spend your time watching videos on YouTube?—but without writing there could be no computer software, no development of the hardware, and much of the content of the Internet remains word based.

Writing in this sense doesn't necessarily mean the use of words but rather expressing information in a way that can be revisited elsewhere in time and space. Originally, before the simplified codes of true writing emerged, the earliest form of "written" communication was by using pictures. The cave paintings of men and animals dating back thirty thousand years or more are not abstract daubing but a means of communicating, whether demonstrating a hunting technique or recording an event. They were fixed in space, slow to produce, and difficult to interpret, but no one can doubt their ability to survive through time.

Over many years, straightforward pictures developed into pictograms. These still featured recognizable images—they are still pictures—but the pictures are usually more stylized, making them quicker to execute and more consistent in appearance. One pictogram will typically represent an object or, more subtly, a concept. It doesn't take a genius to decode a pictogram message that shows fruit lying on the ground, then a pair of arms, then fruit in a basket.

The problem with a system like this is that there are just too many symbols to cope with. A simplification would be to have separate symbols for fruit, basket, and ground and by drawing them in a particular relationship—perhaps with a special linking mark to suggest "on" or "in"—combine those symbols. Now those simple pictograms are evolving into "ideograms"—symbols that can put across a more abstract concept like "on."

This is the stage at which "proto-writing," the immediate ancestor of true writing, is thought to have emerged. Somewhere between nine and six millennia ago, symbols were being used with a degree of visual structure to put across a simple message like the one described. It is very difficult to say when this proto-writing emerged, but many archeologists believe the best early example currently known is that on the Tărtăria tablets, found in the village of that name in central Romania (once part of Transylvania). The clay tablets, a few inches across, feature just such a combination of stylized drawings, symbols, and lines. It's possible that these were purely decorative, but everything about them suggests a message, a clear attempt to communicate information, whether to other humans or to the gods.

The Egyptian system of hieroglyphs is the best-known writing system that takes the next step—still using pictographs and ideographs, but in a much more formalized setting. The big advance here is that the pictures sometimes represent words, sometimes parts of words—another simplification in approach. Although hieroglyphs are the instantly recognizable script of ancient Egypt, they were only used for special purposes. They were slow to produce and not well suited to (for instance) keeping accounts. A second, hieratic, system developed alongside hieroglyphs. Although it is also based on visual symbols, they are significantly more stylized still. Hieratic script appears at first glance more like a modern written one.

The Egyptians weren't the first to use true writing. Another of the regional superpowers, Sumer, had what was probably the first written language, starting with pictograms but gradually translating them to more and more stylized equivalents, ending up with a cuneiform script—one where the characters are built up from wedge-shaped marks (a bit like the side-on view of a tack) made with the end of a stylus.

By around four thousand years ago, writing was spreading like wildfire, whether independently invented or copied from others. The Chinese system dates back to this time, using a large number (around five thousand) of symbols that represent words or parts of words. Our own alphabet has a ragged history before reaching our current written form. The term "alphabet" shows its Greek origins (*alpha* and *beta* being the first two characters of the Greek alphabet), though the letters we use have a more complex history.

The earliest known predecessor of our script is the proto-Canaanite alphabet. Technically it was an *abjad* rather than an alphabet—this is effectively an alphabet without vowels. The vowels are either implied by position or marked using small change marks, like accents. This alphabet was used in the Middle East from around thirty-five hundred years ago and was taken up by the Phoenicians. The symbols were adapted for both Greek and Aramaic lettering. Greek is thought to be the first true alphabet with vowels treated similarly to consonants, developing around three thousand years ago.

Our own Latin or Roman script (at least the uppercase letters) were derived over time from the Greek, and just as the United States and the Internet have spread the use of English today, the Roman Empire spread the use of Latin lettering as their language became a common tongue that would outlast the empire by over one thousand years. Isaac Newton's greatest work, *Philosophiae Naturalis Principia Mathematica,* was written in Latin as late as 1687, while his *Opticks,* first published in 1704, though written in English, was then translated into Latin for a wider audience.

The Roman script we are familiar with—our capital letters minus a few omissions such as *J* and *U*—were the Roman equivalent of hieroglyphs. They were primarily used for carving in stone and making important proclamations. Everyday writing used a different script that is partway between the capitals and modern lowercase, called Roman cursive. Initially the letters varied greatly in size and placement, but over time they became more standardized in scale and more like our current lowercase letters. Originally, capitals and cursive were two distinct schemes—a writer would use one or the other—but gradually over time capitals came to be used for emphasis in the midst of the cursive.

Exactly how capitals were to be used took a lot of time to settle into standardization. In English, for instance, there was a period when they were only used to emphasize new sections like sentences, then a time when, like modern German, they were used on pretty well every noun, before the current balance was settled on. It wasn't until printing came along that the two types of lettering would be called upper- and lowercase, referring to the moveable type that was used in printing until computerized printers became the norm. A page of print was bound together from a collection of individual letters on metal blocks. Capital letters would be kept in higher drawers or cases, while the "minuscule" letters were stored in the lower cases.

Whatever the script used, writing has gone through four distinct phases, each bringing a new urgency and ease of spreading the written word through space. The first was writing on stone, whether painting or carving the letters into a tablet (or simply a handy rock face). This had great potential for permanence but brought with it serious limits of portability. The version most capable of being carried around was clay tablets, where the "stone" was malleable and so could be marked with a stylus, rather than the painfully slow tools like hammer and chisel, and a mistake could be erased and the marking redrawn.

Unusually in the development of a technology, although each new phase of writing medium took over as the frontline means of working, all of the existing phases still continued right through to the present day—we still carve words in stone when we want a short message to last for hundreds of years in a formal setting, though we have other ways of preserving less ceremonial wording. Although clay tablets were a lot more portable than stone, they were still heavy and clumsy, and the use of a stylus didn't allow for the speed of flowing cursive scripts. The second phase was the move to flexible media—vellum and parchment (both made from animal skins) and paper.

This development had three positive effects on the ease of (literally) spreading the word. The medium was lighter and easier to carry, it could be stored more efficiently, in scrolls and in books, and it was much easier to write quickly on with a brush or pen. True cursive script became possible, and with it the speed and flexibility of handwriting. Even more than the tablet of stone, this medium has continued to the present day, and many of us still like to make notes and play around with ideas on paper before committing them to a computer. Some, particularly fiction writers, still write whole books this way.

The only problem with handwriting as a way of spreading the word, rather than just writing it in the first place, is that it is very slow to turn out many copies of a book if each has to be written by hand. In medieval times, copyists in monasteries would spend week after week making painstaking copies of manuscripts. Not only did the process often introduce errors; it also meant that books were inevitably a rare commodity. Reading a book was a luxury reserved for scholars and the rich.

The concept, and our third phase, that would explode the written word out to the wider world was printing. This had been done for many years, making prints from a wooden block on which a negative image was carved (a process still used for illustrations in print right up to late Victorian times). The Chinese were using this process to print pictures on paper and on cloth as far back as the third century A.D., and it was commonly used on paper in Europe by the fifteenth century. But producing a whole book this way was fiddly and it took a huge amount of time to initially set up the blocks, making such block-printed books both expensive and very short. The key breakthrough here was moveable type.

Although this had been used earlier in China, moveable type seems to have been independently invented in Europe by Johannes Gutenberg around 1450. Gutenberg cast individual letters on small metal blocks, which were then fixed together to print a page. The flexibility, accuracy, and relative speed of assembling a page this way once enough type had been cast made it possible to print much longer books and brought down both the cost and speed of printing. Books were still something of a rarity, but the printing press meant that books, newspapers, and other printed material could begin the mass distribution of the written word.

By Victorian times, access to printed material was an essential part of civilization. When Horace Greeley, the founder of the *New York Tribune,* made the journey from New York to San Francisco in the summer of 1859, he noted that even though San Francisco was "hardly yet ten years old," there were between ninety and one hundred periodicals published in California, a third of those being based in San Francisco. When the motion picture pioneer Eadweard Muybridge first settled in California around this time, his job was binding and selling books and furnishing libraries for gentlemen.

The fourth and last phase to be piled on top of the others in our collection of key written media is the electronic representation of text. Although this often simulates the written word on paper, it transforms the way we write and share information. Don't let anyone old enough to be nostalgic about typewriters fool you—they were a pain to use and even worse when you needed to make a change to anything. The electronic production of text, combined with distribution and publication via e-mail and the Web, is as big a breakthrough as moveable type.

So now, with all four phases of written technology in place, we can see the true power of writing in enabling human beings to go far beyond our basic, natural capabilities—in fact, far beyond the capabilities of any natural creature on the planet. Think what I can do, thanks to writing. I can consult the wisdom of someone long dead, purchase something to eat from the other side of the world online, or jot down a reminder on a Post-it note to ensure that I remember to do something important tomorrow. And that's just my direct use. Hardly anything around me that makes me more than an upgrade-free human being would exist were it not for the written word being present in its development. The written word makes it practical to have law and order, science, and literature, to name but three examples. Of course before the existence of writing there was an oral tradition. There were story-tellers, but the difference in capability brought about by writing and its impact on human beings was immense.

Writing is so ubiquitous that it is easy to overlook its significance. Remember that the written word takes the power of language (see page 167)—in itself an extension of our basic biological capabilities—and removes from it the constraints of time and place. It adds in context and illustration. Speech can do a vast amount, but when a topic gets too complicated, writing is necessary to back it up. Writing as a technology was pivotal in pushing our unnatural evolution from a very gradual process to one of exponential acceleration (see page 169). Although writing could be seen as a development of the brain and so fitting better in the following chapter, writing is a tool of power and so fits equally well here.

The written word is immensely powerful, and many of us feel that it has a kind of magic. There is something special about books and bookshops, something very physically satisfying about handling a book. (Equally there's something special about a Web search engine like Google, but that's a different kind of magic.) Of course, as a writer, I would say that books were special because books are what I do, but it's not an uncommon feeling. When written words are combined with a human's practical ability to make things happen, they are of almost limitless power.

With writing came a whole collection of memory aids. It might originally have been obvious to make marks in a cave as a way of preserving information, a place where the marks would be preserved from the weather, or to carve them into a rock with its long-lasting characteristics, but one of the essential features of the brain is portability. Your brain travels with you wherever you go. (This may be self-evident, but it does have implications, and it's one of those facts that are so obvious, we don't always think of the implications.) Caves and rocks are nowhere near as portable as the brain. They are fine to provide a schoolroom in which to note down the fine points in hunting skills to help teach youngsters, but that's no use to you if you are out in the bush and the instructive painting is three miles away.

As we have seen, portability of written material was the next milestone in extending our memories beyond the limits of an individual's brain cells. Initially this seems to have been a matter of using smaller stones rather than rocks, or the malleable "stone" of a clay tablet, but increasingly the medium to carry the word become more flexible with vellum, parchment, and paper, and more practical with the development of the scroll and the book. It was possible at last to have a notebook—a means of recording information on the move—or a prewritten book, to pick up and use information away from a fixed base.

In the twenty-first century, of course, information technology is pervasive. Yet it isn't all high-tech. Many people still make use of a paper notebook for its portability, its practical use anywhere, and its flexibility. For that matter, printed books like this one are still the mainstay of providing in-depth information on a topic or for recreational reading. But computers and the Internet have massively changed our brains' abilities to extend memory and to transfer information from place to place. Communication is a natural ability that we have totally transformed in the last few decades. Landline telephones, cell phones, e-mail, instant messaging, online video—we have vastly extended our communication reach.

In the depths of the ocean, whales can detect their fellows' songs over distances of tens of miles. With the telegraph and then the telephone we were able to communicate directly across a divide of thousands of miles. The cell phone filled in the last part of that particular puzzle, making the communication personal, rather than to a fixed location, just as the portable notepad moved us on from the stone block built into the wall.

Yet for all the power of our technology, when thinking about one important function of the brain—memory—we shouldn't underestimate just how effective our built-in technology, the fundamental capability of the brain, is. Nor should we think that it can't be improved. It is nothing new to realize that memory can work more effectively than it typically does. Ever since the ancient Greeks, human beings have been using explicit techniques to boost the effectiveness of idea retention.

Memory techniques are a particularly subtle enhancement of human capability. Physically they only use what the brain already has, not adding any technology, but they trick the brain into working better than it normally would. To get an appreciation of just how memory techniques work, it's important to have a broad picture of what human memory is. One of the reasons it is essential to look into this is that we now have a very obvious model of memory we are familiar with—computer memory—that is very misleading as a guide to the way the brain stores data.

Computer memory is very straightforward (not technically, admittedly, but in basic function). If you take a memory chip, as you'll find in a computer, a memory stick, or the memory card from a digital camera, it contains a nice structured array of storage, rather like a row of millions or billions of little boxes, one space after another, each holding a bit of information. To store data away, we locate the appropriate bit and set its value. To retrieve our data, we navigate again to the same address in memory (that address is just another number) and read back the value. It's as simple as that.

The human form of memory is very powerful, in some ways much more powerful than computer memory (if much slower), but it's also messier than a computer and more complex in the way it is accessed. Remember, biological systems aren't designed. They can do incredibly powerful things, but they often achieve their goals in a strangely roundabout manner, because of the way they have evolved. It doesn't help that there are several different kinds of memory. Each is primarily *handled* by a different part of the brain, though like many brain functions, memory can't truly be said to take place in any one clear area. Your short-term memory handler is the closest in location to where we'd typically point if asked to indicate where our memory was, up behind the forehead in the part of the brain called the prefrontal cortex.

Short-term memories are strange beasts and may have given rise to the totally spurious claim that goldfish can only remember things for three seconds. Though this makes for great goldfish jokes ("Just because I have a three-second memory, they don't think I'll mind eating the same fish flakes over and over. . . . Oh boy! Fish flakes!") anyone with a garden pond will know that goldfish can learn (for example to come to a particular spot to be fed) and a TV show has demonstrated goldfish learning to navigate through mazes—impossible if they truly had memories that only lasted for three seconds. Perhaps the origin of the myth was some misrepresented piece of information about short-term memory. It's quite possible that a goldfish's short-term memory doesn't function much longer than this.

In principle we can hold things in short-term memory as long as we can concentrate on the things we are trying to remember, but the biggest problem with our short-term memories is that they only have the capacity to store between six and nine items at a time. Try to force too many in, and the earlier items will pop out again. This is why phone numbers are difficult to keep in your mind and why phone companies have traditionally broken up numbers into chunks like area codes that can often be remembered as a single entity. Although identical as a phone number, to the brain (646) 555-4068 looks significantly more manageable than 6465554068.

While we can consciously hold things in short-term memory (in fact, we have to consciously do so to retain the information), we have no conscious control over our long-term memory. This is quite unnerving, when you think about it. We tend to think of ourselves as rational beings, yet here's one of the most important functions of the way that we think, probably the thing that most defines each of us as a person, and we have no conscious control over it. The long-term memory is mediated by an area of the brain called the hippocampus. (This is Greek for "seahorse," as to someone with a very vivid imagination this part of the brain just about looks like one. Personally I think it looks more like a part of the male anatomy.) I say "mediated" because the hippocampus uses other brain cells for storage, but it is responsible for the long-term memory function.

If we want to remember a piece of information—essential if we are going to acquire knowledge and make use of the human ability to go beyond our basic evolutionary status—it's necessary to get the information from short-term memory into long-term memory. This is a gradual process, reflecting the messier "wet" nature of memory in the brain compared with the simplicity and reliability of a computer chip. On the chip, either memory is there or it isn't. In the brain, we can have a vague memory, with just a few outline details, or we can have complex recall, depending on how much a piece of information is reinforced and encouraged to become permanent.

Before considering the techniques that can be used to fix information into memory it is worth outlining a third and very different type of memory. If you have learned to drive, can you remember when you first got behind the wheel of a car? It was probably a painful sight (especially if it was in drivers' ed in front of your peers). Managing to coordinate everything you have to do—controlling the gas pedal and the brake, indicating you're about to turn, using your mirror, watching out for street signs and other cars and pedestrians and traffic signals—it was all too easy to get flustered. But once you've been driving for a few months, everything seems to click into place. It's almost as if you don't have to think in order to drive anymore. (Anyone who drives a car with a stick shift will be particularly aware of how essential it is to get into this state where you don't have to think about the details of what you want to do.)

The reason that driving, and other relatively repetitious activities, become so much easier and smoother with practice is that the instructions on what to do have moved from the long-term memory to the procedural memory. Procedural memory tells you how to *do* something, where short- and long-term memory, both referred to as declarative memory, store information. The procedural memory is located at the "root" of the brain, in the cerebellum and the top of the spinal cord. This tells us that it's a very old form of memory, predating our more sophisticated mental functions, and that it has got fast access to the nervous system, which is essential for this type of activity.

Think of another example of procedural memory at work. If I didn't have any memory of the layout of a computer keyboard, I would have to search through QWERTYUIOP and all those other keys until I found the right location for each letter I wanted to type—the so-called hunt-and-peck technique. To overcome this, I could sit down and memorize the position of each key— but even then I would have to think, "Where's an *L*? . . . Middle row, toward the right-hand end, move along, hit it." After a while, if I used a keyboard often enough, that information would shift into procedural memory and then I wouldn't need to make the effort to remember.

In fact, although I do touch-type, I never explicitly learned where the keys were. Mavis Beacon never came to my aid. I just went into a job—computer programming—where I was sitting at a keyboard all day and gradually picked up the way to touch-type without ever consciously learning the position of the keys. If you asked me where a *B* was, I couldn't tell you. But I can put my finger straight on it without looking at the keyboard. Procedural memory bypasses much of the brain. I just issue the instruction "type a *B*" (in effect) and it happens. This is not only convenient; it's a lot quicker in signaling time than having to send a request to long-term memory to find out where the *B* is, then issue instructions to the nervous system to move the finger and press.

Procedural memory like this isn't our primary concern here. Useful though it is, it doesn't really extend us beyond the basic human model. But getting the right information into long-term memory so it can be accessed, used, and developed at a later time is an essential for adapting our capabilities. Memory techniques enable us to be more effective at getting information from short-term to long-term locations or to make the information more secure in long-term memory. To see how this works, let's take a look at one of the simpler memory techniques, which makes it easy to remember strings of numbers.

In principle, remembering numbers sounds trivial. After all, there are only ten of them to manage—one per finger—how difficult can this be? In fact, this is another of the computer/brain memory differences. While only having ten numbers (or even better two, as computers like to use binary numbers) is an advantage for mechanical storage, to the brain, which was developed to recognize patterns and shapes, numbers are boringly similar. They don't have any great meaning attached to them, so we find them difficult to commit to memory. Yet with a simple technique it's possible to take a new number and push it straight past your short-term memory into longer-term storage with very little effort.

There is one catch to this technique. It works by associating a word with each digit—and you do need to memorize these word/digit links the conventional way, through repetition (this is particularly effective if you speak them aloud and write them down over and again) and revision until, given a number, you can instantly produce the word, and vice versa. Here is one possible set of words for this technique:

One—GUN
Two—SHOE
Three—TREE
Four—DOOR
Five—HIVE
Six—STICKS
Seven—HEAVEN
Eight—WEIGHT
Nine—LINE
Ten (0)—HEN

Note that zero is represented by HEN by linking it to ten. You could represent zero as HERO instead, but it's easier to use HEN, which has a clear visual image attached to it, than HERO in the way these words are to be employed. Now, to remember a number, any number, all I've got to do is build a story around the number, using these key words. I make that story as dramatic and colorful and gross and over-the-top as possible. Let's see that process in action. Imagine I wanted to remember 4372, which, let us suppose, is the PIN for my credit card. (It's not really my PIN, incidentally.)

So, to remember 4372, the magic words that instantly spring

to my mind, because I've got these rhymes really well embedded, are DOOR, TREE, HEAVEN, SHOE. Then I convert these words into a ministry. I imagine myself struggling to carry an immense, ten-foot-high version of my credit card. Suddenly a big door with a huge brass knocker opens in the side of the card. Through the doorway comes a vast purple, bulgy tree, growing at immense speed like something out of a science fiction movie. The disgustingly jellylike tree stretches up higher and higher until the clouds get caught in its branches. Finally, a great big red high-heeled shoe comes down out of the cloud and squashes the tree with a loud squelch.

Notice a couple of details in my example. Things are big, dramatic, and more than a little strange. That all helps fix them in memory. We remember the extraordinary more easily than the ordinary. And though the story is mostly literal (and I could have used any of the "heaven" imagery you are likely to meet in a cartoon or a movie), there's no problem using sky or clouds for HEAVEN as long as you have that well fixed in mind. That story worked for me. The first few times I need to get the number back I have to reconstruct the story, replaying the visuals in my mind as the number is assembled. A door opened in my card (DOOR=4), then a tree came out (TREE=3, so that's 43); it reached up to heaven (HEAVEN=7, 437) and was squashed by a shoe (SHOE=2, 4372). Before long, the number will become so strongly associated with my credit card that I won't need to go through the details of the story anymore. The number will just come into my mind.

This technique gives a real insight into the way that memory works in practice. We find it easier to remember concrete things, rather than something as abstract as a number. We remember stories. Stories fit our natural patterns of remembering much better than the sort of structured information we are used to in computing. Our memories aren't a set of separate, individual addressed locations, like those on a computer chip; they are chains of information. And we navigate those chains using visual imagery, color, and highly memorable key concepts like sex and violence (which I euphemized earlier as "dramatic").

Techniques like this one, building story chains, can be used to remember all sorts of information, not just numbers. If used properly, they become a powerful way to enhance the brain's capabilities. This kind of technique goes back to the ancient Greeks, who used another popular technique based on location. The idea is that if you want to remember a sequence of things, you imagine them located around a room or a house that you know very well. Again we are bringing into the relatively abstract information we are trying to remember context that is more like the "natural" information we have to deal with—spatial location, objects, stories associated with those objects. All this helps to build on our basic ability to memorize and turns our memories into more reliable built-in notepads.

One thing that memory enhancement doesn't help us with is a sense of time. The basic, unmodified human is not very good at measuring time. We can count in a reasonably steady fashion, particularly at rates that can be built around that of the speed of our heartbeat—say anything from about 30 to 120 beats a minute—but we are very poor at identifying how much time has passed while we are active and not watching the clock. It's a subjective truth that when we are busy or interested time seems to pass much quicker than if we are bored or uncomfortable. Our mental measure of the passage of time tends to think that less time has passed than really has if we are enjoying ourselves.

As Albert Einstein famously said, "When a man sits with a pretty girl for an hour, it seems like a minute. But let him sit on a hot stove for a minute—and it's longer than any hour. That's relativity." This was the abstract of a paper Einstein allegedly published in a journal called *Journal of Exothermic Science and Technology,* which described him attempting to undertake the experiment in question. (The film star Paulette Goddard, introduced to Einstein by mutual friend Charlie Chaplin, was the pretty girl in question.) I have only ever seen this paper referred to as a genuine, if humorous, academic contribution, though the way that the initials of the spurious-sounding journal spell out "JEST" might suggest that Einstein made the whole thing up. Funny though this may be, it reflects an underlying truth—time runs away with us when we aren't concentrating on a regular beat.

In prehistory, this wasn't much of a problem. The passage of the Sun through the sky was as good an indication of time as anyone needed. But there would come a point when tracking the passage of time was an essential aid to expanding human capability. All clocks, from the early water clocks and candle clocks, contribute to this mental extension, but the key to truly giving us a personal enhancement in our ability to cope with time was the invention of the watch.

Pocket sundials had been around for some time, but the first pocket watches—small mechanical clocks—seem to have been constructed in the 1500s. The first known maker was Peter Heinlein, a locksmith of Nuremberg, Germany, who is referred to in 1511 as making ingenious clocks that could be carried in the pocket. Initially a novelty, by the nineteenth century the pocket watch was ubiquitous, but a timepiece that is carried in the pocket is relatively impractical to use under pressure, especially when the hands are occupied.

Leather straps were used to bind small pocket watches to the wrist as early as 1880, and by 1901 the British army was using purpose-designed strap-on watches in the Boer War. A few years later true wristwatches were produced. The ability to keep track of the time might not be on a par with language or with the written word in the way it has enhanced our ability to survive and thrive, but there can be no doubt that few modern people can operate effectively without a constant reminder of the time.

The physical aids to the brain we have seen are the equivalent of mechanical extensions to the body like exoskeletons, but as was the case with the body as a whole, it is possible to make more direct impact on the brain by influencing the chemicals that keep the mental processes operating. We have already seen this at the start of the chapter with caffeine, but that is a relatively gentle boost compared with some of the ways that drugs can be used to modify the brain's functioning.

Another drug that alters brain operation and has been in long-term use is nicotine. Tobacco was used by Native Americans, inhaled in high doses to give hallucinogenic effects, but it really took off when it crossed the Atlantic and become the new fad in Europe in the seventeenth century. It should be emphasized that tobacco causes many health issues, but nicotine, without doubt, both acts as a stimulant like caffeine and does give some enhancement to long-term memory. Like caffeine it increases dopamine activity and also boosts production of epinephrine (adrenaline), a hormone that stimulates physical activity in the body.

There is clear evidence that stimulants increase the ability to concentrate. This is why Ritalin, a powerful stimulant, is given to those with attention deficient hyperactivity disorder (ADHD), rather than the more obvious solution of prescribing a sedative to calm a sufferer down. Ritalin is a strong drug with potential side effects, but it does increase attention and memory capabilities in anyone, not just in an ADHD sufferer. According to *NewScientist* magazine, up to 10 percent of U.S. university students take Ritalin or other prescription stimulants as so-called smart drugs to boost their attention and concentration.

There is also a big rise in the use of modafinil (trademarked as Provigil) by students. This is a wakefulness promoter that was devised to help narcoleptics, people who suffer from a condition that can make them suddenly and dangerously drowsy, to help them avoid losing their alertness. Modafinil has a more powerful effect of keeping the user awake than caffeine, with less of the jittery feeling that is associated with too much coffee. Modafinil isn't limited to keeping you awake longer. Tests by Danielle Turner at the University of Cambridge in England have used modafinil on subjects undertaking an intelligence test involving manipulation of shapes and found significant improvement in performance, particularly on more complex versions of the task.

Unfortunately, all the stimulant drugs have issues, both in being habit-forming and in having potentially harmful side effects, particularly when taken to excess. There needs to be more development work on this kind of medication—but it is on its way, particularly inspired by modafinil. A colleague of Turner's, Barbara Sahakian, has described modafinil as the first true smart drug. There will be many more. Dozens of drugs specifically intended to improve cognitive function are under development.

Researchers have pinpointed at least fifteen different routes of chemical attack in the brain that could improve mental processes. Those involved in the field say it's not a matter of "if" but "when" there are effective over-the-counter medications to improve thinking capability.

Not everyone is comfortable with taking medications to enhance their mental capabilities. Helping the brain function more effectively by eating and drinking the right things seems a much more natural approach than taking drugs, even though at the chemical level there is no real difference. Traditionally fish was considered "brain food," with some justification, but now there is a whole range of "smart" food and drinks whose makers claim that their products will help the brain function more effectively.

The aim of taking nutrients to help the brain is usually to assist in the production of one or more of the chemicals used by the brain in thought and memory, hence hopefully making an improvement to your thinking power. Sometimes the effectiveness of the nutrient is in doubt, because the reasoning behind the product relies on the "hair product syndrome." Many hair products are advertised as containing something that has positive associations with health—fruit, for example—in the assumption that it will make for "healthier" hair. Since hair is dead, it can't be truly healthy (though it can be structurally modified), and there is no reason to assume that something that makes the body healthier when eaten will have a beneficial effect on hair when rubbed onto it. The parallel with nutrients is that it is not always the case that eating or drinking a substance used to produce one of the brain's functional chemicals will make the brain any more effective. But in some circumstances there is good evidence of benefit.

One example is the neurotransmitter acetylcholine, which is important in memory formation. The body produces acetylcholine from a simpler chemical called choline, found in fish, or from lecithin (which is used to produce choline first), found in some seed oils. There is evidence from experiments at MIT that students taking a regular dose of choline or lecithin had an improved ability to remember lists of words.

Boosting intake of choline or lecithin does seem to have beneficial effects on the ability to form new memories, particularly in older people who have less efficient acetylcholine production than youngsters. The nutrients can be taken in through supplements, smart drinks that contain the chemicals, or a diet with more products rich in choline or lecithin. For example, soybeans, liver, egg, peanuts, pulses like peas and beans, yeast products, and green vegetables all boost levels of these nutrients.

Other contents of smart drinks and supplements over and above the familiar vitamins include amino acids like phenylalanine and glutamine, both important in brain function, and omega-3 oil, which is an essential fatty acid ("essential" meaning that it is one that we have to consume because the body doesn't produce it) and a nutrient that we are often low on. (The other essential fatty acid, omega-6, we tend to have plenty of.) Omega-3 is important in preventing heart disease and strokes but also contributes to brain function.

The all-round usefulness of omega-3 is probably why fish has got the reputation of being brain food. Fish, and particularly oily fish such as salmon, mackerel, and tuna, is about the best natural source of omega-3. Apart from eating the fish itself (which isn't recommended more than about twice a week, as unfortunately many oily fish contain contaminants such as mercury), there are plenty of fish oil supplements available, and omega-3 is available in a wide range of products from smart drinks to fortified margarines and olive oil.

When it comes to the actual formation of memory in the brain, one protein, cyclic AMP response element binding (CREB for short) protein, has a big role to play, as it is used by the brain in the construction of synapses. Synapses are the tiny junctions between pairs of brain cells and also form the links between the nervous system and other parts of the body. Each neuron in your brain—you have around 100 billion—is connected to anything between a handful and 1,000 other cells. In children, there are around 10,000 billion of these synapses, falling to something like 1,000 billion as we get older.

This dropping off, incidentally, isn't the same as the old idea that brain cells die off gradually through our life and are never replaced. We now know that brain cells do regenerate throughout life, but the number of connections in the brain does gradually reduce. We also know that memory is dependent on the synapses, and as Nobel Prize winner Eric Kandel discovered in his work with giant sea slugs, the protein CREB appears to make it easier for memories to form. Kandel, who escaped the Nazis in Vienna as a boy to become a top scientist as a U.S. citizen, has spent his whole working life exploring the nature of memory at the level of individual cells in the brain. After working for many years with giant slugs, he went on to use mice, raising the levels of CREB in their brains—the result was to produce mice with memories that were twice as good as untreated animals had.

It might seem impossible to tell how good a mouse's memory is. You can't ask it who won the last Superbowl or who is currently president of the United States. At one time, mazes were used as a test of memory. Mice had to find their way through a maze to get to a tidbit of food. The faster they learned the maze route, the faster the memory was assumed to have formed. But it has since been shown that mazes aren't a great way of establishing memory levels, because the skills needed to negotiate a maze are quite different from pure remembering of information—there is too much procedural memory involved.

In Kandel's experiment, the mice were placed in the middle of a brightly lit circular table with holes around the edge. Mice don't like to be exposed to bright light or to be in the center of an open space, feeling in danger. They try to find a bolt-hole to get out of sight—but only one of the holes around the edge of the table was a way to escape: the rest were dummies. Initially the mice would randomly try holes, then they would begin to take a more systematic approach, but eventually they remembered a useful fact. Markings on the walls around the table showed which hole was the escape route. Although the position of the escape route changed from session to session, the marking was moved with the hole. This way, mice that remembered the connection of marking and hole would escape more quickly than those guessing at random or working systematically around the table.

The result of these experiments, and parallel work elsewhere, is the development of drugs, not yet on the market at the time of writing, that can be used to help with memory impairment, such as that caused by Alzheimer's. But the sick aren't the prime market for these drugs. Drug companies want the biggest customer base, and there are more "normal" people than patients with memory impairment. If the drugs can be shown to work effectively and safely, the real dream of the drug companies is to have a pill that, taken regularly, can be used to boost memory function in ordinary, healthy human beings.

Enhancing the brain's capabilities directly is not only amenable to influence by chemical means. It is thought that magnetic fields can be used to enhance memory and to improve general brain function. Initially there was a lot of suspicion about the technique of transcranial magnetic stimulation, which uses powerful electromagnets to influence the brain. It seemed all too similar a concept to the eighteenth-century fad of Mesmerism or "animal magnetism" (referring to the "animus" or spirit, not animals per se), which claimed with no scientific basis to provide medical cures by stimulating the "magnetic field" that was believed to surround human beings like an aura.

However, powerful magnetic coils to stimulate the brain have been used experimentally in the last few years to treat brain disorders and to help with recovery from strokes. The strong magnetic field induces electrical currents in the brain, which kick various neurons into action. Fortunato Battaglia and his team at the City University of New York have shown that using this transcranial magnetic stimulation on mice increased the action called long-term potentiation that is used to store memories away.

The treatment also increased the levels of stem cells in a region of the brain called the dentate gyrus hippocampus. These cells continue to divide throughout our lives, and research at Johns Hopkins University School of Medicine in Baltimore, Maryland, has shown there seems to be a connection between these new cells and the ease with which we can store away new memories. More research is still required, but it does seem that appropriate focused magnetic treatment will be able to help hold back the impact of memory-impairing diseases like Alzheimer's and may help any of us improve our memory formation.

Impressive though the results of external stimulation are, not everyone thinks that it is enough to stay outside the brain when trying to enhance its functions. Hands-on brain surgery is also an option to give the brain a boost. Experiments are under way to enable implants to communicate directly with the hippocampus, that (roughly) seahorse-shaped segment of the brain that plays a major role in handling long-term memories.

In 2006 a team at the University of Southern California led by Theodore W. Berger replaced part of a slice of a rat's hippocampus with a chip, which was able to interact with the brain segment, emulating neurons, successfully processing the signals that are transmitted through the hippocampus. The chip had been under development for several years, following painstaking work on hippocampus cells, stimulating them again and again millions of times and recording their responses. This was necessary as we don't understand how the hippocampus processes memories, so the hippocampus cells were treated as a black box, and the chip mimics the response of the cells to different stimuli.

The USC team's follow-up experiments, started in early 2007, involved taking a step back from the chip itself, using a computer to simulate it, as the specially built chips are very expensive to make. But the intention is to move from working with extracted sections of brain to communicating with a living rat's brain. To take the experiments to the next stage would require embedding the chip in a rat, which would mean avoiding the rejection mechanisms of the body and also getting around the brain's ability to reorganize connections around problems, which it may well regard the chip as being. From rats, it's expected that the experimentation will move on to monkeys and eventually humans. (The chip would not be implanted in the brain but sited externally, communicating with the hippocampus through electrodes.) Like any work on the brain, this is a nontrivial task—back in 2003, when the chip was first being constructed, it was expected to be trialed on live rats within six months. In practice it has taken four years.

The move from rats to humans itself leads to problems, over and above the obvious issues of risk to the subjects, and not being sure just what the experimental subject would experience. The implant has to be able to model the action of the neurons in the brain, but there is currently no way to do this without disrupting brain function. We have no way to safely scan brain signals at the level of individual neurons without being intrusive, which has meant that researchers have suggested they may have to use models of monkey neurons for human trials. However, in the timescales involved it is quite possible that nonintrusive scanning at this level of detail will be available before human trials could be conceived. Berger's Web site suggests human trials could be possible in "the next five years," but this seems as optimistic now as the "six months to rat trials" estimate was in 2003.

Memory enhancement chips like this were designed to repair brain damage, enabling memory function to be restored when it is failing, but they could also conceivably be used to boost brain activity. Although the brain is an incredible, vastly complex structure, neural mechanisms are very slow compared with electronics—a memory-boosting chip could in principle make memory faster and more efficient.

As we will see in the next chapter, we are already acting directly on the brain, implanting electrodes to deal with serious medical conditions. But it is much harder to justify operating on the brain—with all the attendant risks that this brings—where the only intention is to provide enhancement to a healthy person. Whether we want to think better, to remember better, or simply to get more out of video games and the next generation of person-to-person communication, these are not necessarily good enough reasons to open up our most essential organ to the knife.

Some futurologists love the idea of the wired human, the person with the socket in his or her skull to jack into the electronic world and expand his or her mind. This was seen in dramatic form in the Matrix movies and variants have cropped up in countless books before and since. However appealing the benefits, though, it seems hard to believe, with our natural squeamishness about the brain, that many of us would allow ourselves to be tampered with at this level just for fun. One thing is certain—if wired connections to the brain ever did become a commonplace reality, those frightening-looking connections in the head we see in the movies would be a nonstarter (especially the huge, unsubtle sockets of the Matrix films).

The biggest problem with introducing electrodes into the brain, apart from the nontrivial risk of damage during implantation, is that the point at which the wire (or socket) passes through the scalp and through the skull is a potential source of infection that would constantly put the brain at risk. It is inevitable that should brain interfaces become everyday, they will be located wholly under the skin, using noncontact methods of communication, like the Radio Frequency IDentifier (RFID) tags frequently used now in stock control. However much we can move away from the unsightly and dangerous sockets in the head, though, greater hope must be held out for developments in external, nonintrusive electronic brain interfaces that get away entirely from the need to cut into the skull.

Many attempts have been made to use variants on electroencephalographs (EEGs) to provide control from the brain with nothing more than a set of electrodes that rest on the scalp. The subject usually wears a plastic cap, which positions electrodes around the skull. As the neurons in the brain fire, tiny electrical charges are generated, which the EEG picks up. Unfortunately, there are so many neurons in the brain that it is currently impractical for an EEG to detect anything other than the average output across millions of different cells. The result, compared with the precision that direct electrodes can provide, is a blurry, limited control that is easily misled by other brain activity. Even to gain this limited control, to be at all effective, EEG-based controls can take months of training.

Further down the R & D chain for the moment, but still something that would be ideal for those with impaired senses, is the reverse of an EEG helmet, a device that can stimulate neurons in the brain without implanting electrodes. Although there are techniques to produce a current without direct physical contact, like the induction units used to charge electric toothbrushes, there isn't at this time a way to focus this induced current. A stimulus version of the EEG would need to be able to target small numbers of neurons or even single cells to deliver a signal. While there isn't a direct technical solution to this problem at the moment, there is every possibility that there will be a technology developed to enable this.

The limited capabilities of EEG helmets haven't prevented

control devices based on the EEG mechanism from being produced. There have been a number of attempts to sell game controllers based on this technology, most recently Emotiv Systems' Epoc, which was demonstrated in early 2007. The ideal, of course, would be to have the accuracy of the brain implant combined with the noninvasive nature of external EEG sensors. It is possible that with enough sensors and enough computing power this can be achieved, providing pinpoint accurate 3-D scanning of neural activity from an external device.

The chances are, such a device won't be based on the same mechanism as EEG, but there are increasingly many other ways of detecting brain activity that hold out a hope for the future. Quantum effects, magnetic activity, and optical response to wavelengths that can pass through the skull are among the many options that have been considered or that are under trial. This really does seem to be the sort of application where the answer isn't "if" but "when."

We tend to think of the brain as the lump of matter that resembles an enormous gray walnut in our skull, but the exact line between brain and body isn't always easy to define. The nervous system is more closely related to the brain than it is to the rest of the body, and this system that stretches from head to toe is little more than an extension of the brain. You could say that, for instance, the optic nerve is a system that joins onto the brain, carrying information from the retina of the eye, but equally you could think of the brain as extending down the optic nerve all the way to the retina.

This is not just playing with words. There is a lot of preprocessing that goes on in the eye before information is sent to the brain. There are many more sensory cells in the eye itself than there are fibers in the optic nerve—the signals from the sensors (rods and cones in the eye) are collated before being fired up to the brain. Effectively, there's a part of your brain that resides at the back of your eye. For this reason, it seemed best to include enhancement to our senses here in the brain chapter, rather than in the chapter on bodily strength.

Sensory enhancement tends to come on two separate levels— the direct physical and the indirect virtual enhancement. We have a natural tendency to rate the physical as a more "real" experience than is the virtual, although at the pure physical level this distinction is hazy. Let's say there's an eclipse of the Sun only visible in Florida. Just how real our experience feels while watching the eclipse fits on a spectrum, depending on how the light from the Sun gets to our eyes. I could travel to Florida and watch the Sun disappear behind the Moon directly through eclipse glasses. (Important warning: never look directly at the Sun, even during an eclipse, without appropriate protection. It can seriously damage your eyesight.)

Or, still in Florida, I could project the eclipse through a telescope onto a piece of card and see what happens by watching that card. (Even stronger warning: never, never look through a telescope toward the Sun under any circumstances; it will lead to certain blindness.) Alternatively I could follow the progress of the eclipse live on TV and save myself the trouble of traveling. Or, even more indirectly (since I'm going to be in a meeting when the eclipse takes place), I could record the TV show and watch it when it's convenient for me to see it.

Each of the options takes me one stage further away from what I would regard as a direct experience. In the first case, the light generated by the Sun travels across space, through my glasses, and straight into my eyes. In the second, that same light passes through an optical instrument, hits a piece of card, and is reflected into my eyes. In principle I am looking at the same light as when I look directly. (In principle only—the reflection process usually involves the photons of light being absorbed and new photons being emitted.) Yet the experience seems more remote because I am not looking at the object itself (the Sun); I am looking at an image of that object.

When I watch the eclipse on TV I am even further removed. I am looking at an image that is generated by electrical signals that were produced by an electromagnetic wave that was generated by other electrical signals produced by receptors that received the light from the Sun. I am more emotionally detached because of the indirectness and because I'm not "there." And in the final case, I've even lost the sense of being there in time. It's a frozen piece of history, not something that is happening in the "now."

Direct sensory enhancement has more of that feeling of being linked to the experience. A telescope, for example, is a powerful direct enhancement of the visual sense. When I look through the telescope, I feel as if I am still seeing the same thing as I would see with my eyes, but I see it as if I had much more powerful vision. In fact, the chances are that I am no longer seeing the photons of light that originally came from the object I want to view—they will have been absorbed and new photons reemitted inside the telescope—but there's still that feeling of immediateness that is quite different from watching a recording of the view through the same telescope on a TV or computer (the view, incidentally, that most professional astronomers now have—observations are very rarely made by eye anymore).

The desire to see farther than the eye is capable, whether into the far distance of space or into the world of the very small, goes back a long way. Many people think of Galileo as the inventor of the telescope, but in the thirteenth century the early scientist and friar Roger Bacon (see page 32) was already speculating that it was possible to use lenses this way. Of all the sciences, Bacon was particularly fascinated by optics, and was certain that they might be used "so that the most distant objects appear near at hand and vice versa." He goes on to say that using lenses "[w]e may read the smallest letters at an incredible distance, we may see objects however small they may be, and we may cause the stars to appear wherever we wish."

This isn't the result of some vague thought that because lenses distort things they might make it possible to see somewhere distant. Bacon describes the basic optical principle behind the telescope and microscope: "For we can so shape transparent bodies, and arrange them in such a way with respect to our sight and objects of vision that the rays will be refracted and bent in any direction that we desire, and under any angle we wish we shall see the object near or at a distance."

Bacon may even have built a crude telescope, but if he did, no record of it remains. Before Galileo came on the scene, though, others would dabble with extending the capabilities of human eyesight. Leonardo da Vinci is known to have experimented with lenses, and it is now thought that a practical telescope was demonstrated by the father-and-son team of Leonard and Thomas Digges, who were working in England at the time of the first Queen Elizabeth, around forty years before Galileo's work. Shortly after there was an optical boom in the spectacle-making towns of Holland, resulting in the early examples of practical telescopes and microscopes that would inspire Galileo and all the later developers of optical instruments.

Optics provides a direct method of extending one of the functions of the brain, by steering, collecting, and focusing photons to enhance images and make things appear bigger than they would to the eye. They also allow for more precision. The human eye is very sensitive, but our eye/brain combination allows for a whole range of misunderstandings that we call optical illusions. One of the most overlooked of these is the apparent size of the Moon. We all know how the Moon can look much larger than normal on some occasions (if not as big as Hollywood tends to portray it). In fact, we almost always see it bigger than it should appear to be. The true apparent size of the Moon is about the same as the size of the hole in a piece of punched paper held at arm's length. You can try this out with the full moon.

Yet the direct manipulation of photons by optics is only a minor part of the ways that we have managed to upgrade the sensory extensions to the brain. When Roger Bacon described using lenses to see at a distance he imagined that it would enable someone to see into a walled city, presumably using mirrors in some way. In practice a telescope doesn't let you see through city walls—but we have a device that can do this and much more. In fact, it enables us to see anywhere else on the planet and even out in space—it's TV. Like many millions of others of a certain age, I sat watching the TV on July 20, 1969, as a fuzzy image of Neil Armstrong descended the ladder from the lunar lander onto the Moon's surface. Thanks to this technology, I was looking at something happening on a different celestial body.

This might seem to be cheating: TV requires that there's a camera in place and all the networks in between to bring the information to me, but that doesn't stop it being, in effect, a stunning extension of our senses. I can sit at home, staring at a box, and see what's happening in Washington or London, Iraq or Afghanistan, with far better precision that was ever claimed for a crystal ball. Although the technology makes my view more indirect, that's just the mechanism. The impact is the same as extending my senses (I can hear as well as see these places) out all over the world.

The Internet has already been mentioned—as with TV, this provides an extension to our physical sensory inputs. Where historically I would have had to physically travel to a library and haul books or journals off a shelf to read information with my eyes, now I can often connect through this network—again, the feeling is one of extending my senses like the tentacles of some far-reaching creature—to this information without ever leaving my desk.

A huge span of technological advances—the telegraph, radio and TV, the telephone, the Internet, spy satellites, GPS and Google Earth, sound amplifiers and X-rays, and more—have extended the basic capabilities of our senses, stretching them out in a filigree network of sensation to the farthest parts of the planet and even into space, giving us the ability to know exactly where we are located (a sense we don't naturally possess), enhancing what we are biologically capable of, and bringing the impossible into reality, like using X-rays to see through objects that light can't penetrate.

To date these enhancements have taken place external to the body, feeding in through our standard communication channels, but in the future we could link directly to the brain to provide a whole new level of upgrade. As we will see in the next chapter, a considerable amount of work has been done on feeding information into the brain as if it came from the eyes or the ears, to help overcome blindness and deafness. At the same time, other scientists have been looking at relaying the messages the brain receives to the outside world.

In 1999 a team at Harvard led by Garrett B. Stanley took signals from a cat's retina and used a computer to convert these electrical impulses into a TV picture. They were able to present a crude representation of what the cat saw on the screen. Other researchers are going even further. A team at Vanderbilt University under Jon Kaas has been investigating direct brain-to-brain communication. By implanting electrodes in the brains of macaque monkeys, in the section of the brain responsible for hearing, Kaas and his team have been recording the signals produced by hearing.

With a good map between what is heard and the stimulation received in the brain, the hope is that it will eventually be possible to produce the effects of hearing those sounds by replaying the stimuli into the brain. With similar studies on the brain cells that fire when sounds are generated, it should in principle be possible for two individuals to communicate brain-to-brain via an electronic link. Not exactly telepathy, but something very close (see page 207 for a crude alternative version of this that has already been demonstrated).

An eight-hundred-thousand-dollar grant to Kaas is only part of the investment that DARPA (whose predecessor, ARPA, was responsible for the ARPANET, the precursor to the Internet) has made in electronic interfaces to the brain. It has invested millions in other projects, all aimed at providing a better understanding of how we can interact directly with the brain, with the long-term goal of being able to control and get feedback from military craft, either to improve response times or to remove the human to safety entirely, controlling the craft remotely but with just as good a sensory input as if the pilot or driver were present in the hostile environment.

Many of the attempts to link the brain to electronics do so at the level of the brain itself, but others use our body's natural communication network to the brain, the nervous system. After all, this is how the brain habitually receives input and translates its wishes into actions, so it seems a sensible way to link into the brain. What's more, there's the advantage that the surgical requirements are less risky when, for example, connecting a chip to the nervous system in the arm, rather than by drilling holes in the head.

Since the late 1990s, Professor Kevin Warwick of the University of Reading in England has been experimenting with a range of implants under his skin that have, rather dramatically, been referred to as the first steps on the way to a cyborg. The term "cyborg," a contraction of "cybernetic organism," has been around a surprisingly long time. It was used in *The New York Times* as early as 1960 and has frequently appeared in science fiction, usually referring to something implacable and terrible, as in the "Borg" that first appeared in *Star Trek: The Next Generation.* In principle any built-in electronic enhancement makes the individual a cyborg—so Warwick's description of himself is correct, though most users of the term tend to imagine a much more radical electronic/mechanical component in the mix.

Warwick's first implant was little more than the type of RFID chip now used to monitor inventories from books to clothing. On the inside of his upper left arm, a small chip in a cylindrical container was implanted between his skin and the muscle. For nine days it enabled Warwick to be recognized by various electronic devices around the offices of the Department of Cybernetics where he works.

As Warwick walked around the building, computerized voice boxes would say hello, doors would open, and lights were switched on and off by his presence. Of course, this was doing nothing that isn't possible with a simple security tag in the pocket or an ID badge, but the significance was the remote communication with an embedded chip. Warwick was to go significantly further in 2002. A second chip was implanted, this time with a one-hundred-electrode array that was connected to Warwick's median nerve fibers, below the elbow joint of his left arm. With this connection, Warwick was able to have some control over both an electric wheelchair and an artificial hand.

Perhaps most interesting of all, Warwick's wife, Irena, was also given a (less complex) implant. With this, she was able to send an artificial sensation to Warwick. A command from her brain activated her implant, which generated a signal. This was then translated into a signal sent to Warwick's implant, which finally generated a sensation in Warwick's brain. A communication from Irena's brain was sent to Warwick's brain by electronically extending and connecting their nervous systems. Warwick's implant was also stimulated to give a sensation by input from an ultrasonic sensor—in effect, he was "hearing" ultrasonics—and he was able to extend his nervous system over thousands of miles, connecting through the Internet to control a robot arm the other side of the Atlantic.

Warwick has frequently inspired controversy because he is very media savvy and exposes his work more directly to media scrutiny than most scientists (though at the time of writing little has been heard from him in a couple of years). He may be a maverick, but he is not a charlatan. Certainly his work to date has been more "proof of concept" than anything practical and he has not been backward at getting his name in the press. Yet his work does offer significant advantages of relative safety over direct brain cybernetic connections, and there are some real hopes from his demonstrations that electronic extension of some aspects of human capability could be performed this way.

Pulling together all the possibilities that we've seen in this chapter provides what is both the most exhilarating and the most frightening glimpse of an enhanced future, provided we can either get external scanning caps and stimulators working with precision or come to accept the practicality of an intrusive brain implant. If this is then combined with a sophisticated computer, the transformation of the human experience would be stunning.

Imagine entertainment that can stream directly into all your senses. Communication with your partner that isn't limited to voice and body language but can express intimate feeling directly. Imagine no longer needing to go and look up information on the Internet but being able to ask for it in your mind and it's there. The idea in the Matrix movies that characters could gain skills and information when an operator plugged in the appropriate memory disc is about as modern as the idea that a telephone call relies on operators connecting physical wires in exchanges. A real-life Trinity would learn to fly a helicopter in this imagined future simply by thinking about the need.

It's overwhelming. The more you explore the implications, the more dramatic the possibilities become. Even so, it wouldn't necessarily be all rosy. Let's look at some of those highlights again, with a more skeptical frame of mind. Yes, it would be great if you could experience all the sensation of being at a great rock concert by thinking about it—but it's all too possible that advertisers would find a way to project their irritating messages into your brain. And if you think you've got problems with spam and viruses on your PC, are you ready for brainspam and to have hackers attempting to crash your cortex?

Similarly it's true that in a perfect situation it would be amazing to share the feelings of pure love in a relationship. But equally this would imply a world where the white lie was no longer possible. It might prove difficult not to communicate what you *really* feel when your partner says, "What do you think of my new hairstyle?" And that would open up a whole can of worms that not everyone wants to experience.

This, then, is probably another argument for using the removable interface, rather than being permanently wired in. A radical brain interface to deliver this sort of capability is not a few years but decades away. Even so, there is nothing in this view to the future that is excluded by what we know of the science or of the brain.

At the moment, the only activity where wiring is made directly to the brain is in the medical field. Where this chapter has focused on methods of enhancing the brain's capabilities when it is working well, both the brain and body often also need fixing. Unfortunately, for thousands of years those who put themselves into the hands of doctors and surgeons have discovered that the will to make things better isn't enough to guarantee success.

6.
Body Shop

The only parts left of my original body are my elbows.

—Phyllis Diller

A simple tool is easy to repair, but human bodies are anything but simple. We all have aspects of our bodies that would benefit from fixing, if it were easy—but for thousands of years our attempts to make things better have resulted in as much damage as they have benefit. The medical profession may have a long history, but much of that history has been very undistinguished indeed.

Some upgrades to deal with faults in the body *were* simple and delivered value from the earliest times. Using a walking stick goes back well into prehistory. Eyeglasses (sometimes attributed to Roger Bacon, though there is no contemporary evidence for this) were developed no later than the Middle Ages. And variants of the wheelchair have been around as long as we have had wheeled vehicles. Prosthetics, too, replacing lost limbs, have been around for a long time. Wood and metal have been in use for a couple of thousand years (there is an example of an artificial leg in an Italian tomb dating back to 300 B.C.), though many fewer patients survived amputation in the early days without anesthetic or antiseptic to have need of an artificial limb.

To begin with, replacement legs were little more than strap-on crutches, though since the twentieth century there have been attempts to make artificial limbs and other prosthetics more realistic and to give them more function, making them a better replacement for what has been lost. As we will see later (see page 231), prosthetic work is taking big steps forward in being able to interface with the nervous system to give a natural control of the action of the prosthetic. At the other extreme, the simplest mechanical replacements are now commonplace—we regularly replace failing hip and knee joints with artificial equivalents, enabling older people to have a more normal, active life.

However, to do much more than is possible by propping up the failing body took the development of effective medicine. Broadly, ancient medicine divides into three strands—spiritual, chemical, and surgical.

Spiritual medicine was devised largely on the basis of attempting to modify or appease spiritual forces, either within the patient or coming from the outside. Although conceived as medical, many of the cures following a spiritual line were in practice an attempt to apply magic. For example, the cure could involve some form of sympathetic magic—making use of a root or an object that bears a visual resemblance to the affected part of the body.

Arguably the homeopathic principle, which aims to cure by giving highly diluted doses of a poison that causes similar symptoms to the original disease, is another spiritual medical principle. Note, by the way, this does not mean that homeopathy does not work (though the best evidence currently suggests this treatment works by the placebo effect). Some natural cures that certainly do work were stumbled on by accident as a result of magical or spiritual reasoning. It's possible to have an effective medication that was originally selected for a totally illogical reason.

When herbal remedies, among the earliest known medications, were first used, they, too, were spiritual cures, even though the actual effect they produced (some beneficial, some anything but) was chemical. It was often specified that an herb should be picked at a time of astrological significance—for instance at the full moon—if it were to have effect, emphasizing this spiritual component of the herbal treatment.

Astrology, which dates back at least five thousand years in the Middle East and was independently devised in China around four thousand years ago, was originally a significant component of the spiritual side of medicine. The practice, long dismissed by science, depended on the idea that human beings could be influenced by the heavenly bodies. This type of astrology was not an attempt to foretell the future, like a modern newspaper horoscope, but an indication of how the Sun, the Moon, planets, and stars would influence an individual's physique in the present.

This aspect of astrology, though still without any scientific basis, was more logical than the fortune-telling variety. It was known, for instance, that people felt better when the Sun was shining, so why shouldn't all the other heavenly bodies have an influence on our spirits? Astrology would continue to have a significant influence on medicine until medieval times.

Spiritual medicine was also very obvious in many cultures' early medical practices, from the ancient Egyptians to the Native Americans, and has survived most strongly in the present day in

Chinese traditional medicine. Often rituals, incantations, and incense would be used in an attempt to drive out dangerous spirits or to have an influence on the spirit of a suffering individual. Yet few early civilizations kept their medicine on a spiritual level alone. Many were also aware of the benefits of surgery and of some chemical treatments.

In a sense, surgery has the advantage of being the most obvious approach to medicine—it's more like carpentry than the subtle diagnostic requirement for effective chemical treatment. Something is broken and it needs fixing. Surgical manuals dating back over three and a half thousand years have survived from ancient Egypt, while Indian medicine was already detailing surgical procedures at least two thousand years ago. In Europe, surgery's hands-on nature meant that it was considered a task for a craftsman rather than a professional such as a physician. Surgery was carried out by barbers, not doctors. (This is why the traditional barber's pole carries a red stripe, for blood, and a white one, for bandages.) Even today in some countries surgeons are called Mister rather than Doctor because they originally weren't medical professionals, though the "Mister" label has gained an inverted cachet.

The Arab physician Ala-al-din abu Al-Hassan Ali ibn Abi-Hazm al-Qarshi al-Dimashqi (known as Ibn al-Nafis) made a reasonably accurate description of the circulation of the blood as far back as the thirteenth century, but Western chemical medicine was based on an incorrect model of how the body functioned all the way from ancient Greek times to the nineteenth century. (The English doctor William Harvey independently discovered the circulation of the blood in 1616, but this had little impact on the diagnostic approach used at the time.)

The model of how the body worked—unsurprisingly wrong, as it was based on pure philosophical theorizing rather than any examination of the innards of human beings (or animals)—was the idea of Hippocrates, a Greek philosopher probably born around 460 to 450 B.C. on the island of Kos, though much biographical data from this period is highly suspect. Hippocrates seems to have been inspired by the four-elements concept of his contemporary Empedocles.

Empedocles believed that everything was made up of earth, water, air, and fire. The concept was wildly wrong as an identification of the elements, though it does correspond roughly to the first four states of matter—solid, liquid, gas, and plasma—and it is understandable when you, for instance, burn a piece of wood and see moisture coming off with flame and smoke, leaving ash. Apart from leaving us with his famous oath, Hippocrates dreamed up a parallel medical structure of four humors, each corresponding to one of Empedocles' elements. For earth there was black bile, for water there was phlegm, for air there was blood, and for fire came yellow bile. The structure also neatly matched up with the four seasons.

These four substances were all fluids that had been observed either oozing from orifices or leaking out when the body was damaged. Hippocrates' idea was that these four liquids were sloshing around in the body in approximate balance. But should one or another of the humors become dominant, the body would go out of kilter and the result was illness. What we eat and how much exercise we take, which clearly influence how we feel, were thought to change levels of the different humors.

The humor model was reinforced as the definitive description of the body's workings by Galen, a Greek physician from the time of the Roman Empire, born around A.D. 129. Galen was a daring surgeon (probably irresponsibly so), performing both eye and brain surgery, but these efforts would have much less impact on medical history than his support for and development of the humor theory. Unfortunately, this stress on the humors led to medical practice that didn't just fail to help make patients better but actively made them unwell.

Many problems, for example, were put down to an excess of blood. Anything from headache to fever was thought to be a result of an imbalance that could be cured by removing the excess blood—a process known as bloodletting. Although I have classified this as part of the chemical treatment, as it was based on the humor model, the bloodletting itself was usually performed by the barbers who took the role of surgeons. It was undertaken either by making an incision or by using leeches to suck the blood out.

Bloodletting inevitably left the patient weaker—in fact, a common way of deciding when the right amount of blood had been taken was to keep it flowing until the patient fainted—and so the treatment had the direct effect of reducing a patient's ability to fight off any infection. Similarly, other treatments to remove unwanted humors—making the patient sick or inducing diarrhea, for example—did much more harm than good.

Probably the most effective medication undertaken in medieval times was herbal medicine, which gradually escaped from its spiritual background to become chemical as a result of a pragmatic observation of what worked. Although some herbal remedies had no real effect (and others verged on the deadly), it was certainly true that the extracts of some herbs would help relieve pain and provide other benefits. Both aspirin and quinine were first discovered in a natural form in the bark of trees, with the willow bark that produced the painkiller recorded as far back as Hippocrates' time.

In the seventeenth and eighteenth centuries, apothecaries who practiced herbal medicine, notably Nicholas Culpeper, who worked in seventeenth-century London, were constantly battling with the physicians, a tightly organized cadre who were reluctant to allow outsiders to gain access to their patients (and their patients' purses). But Culpeper's move of herbal wisdom from hearsay and old wives' tales to a structured, semiscientific approach was an important step in the acceptance of chemical medication. Even so, in Culpeper's day there was still a significant spiritual component to the herbal, with the expectation that how and when you picked the herb could influence its effectiveness.

So with apothecaries still influenced by the spiritual and doubted by the rest of the medical profession, with surgeons who were barbers on their days off, and with physicians who based their treatment on a totally fictional model of the way the body worked, it's not entirely surprising that many people before the nineteenth century were better off if they didn't seek medical help. At this stage much of medicine was not so much a step forward beyond our natural state as a regression.

Even by the 1950s a lot of medicine was, frankly, guesswork and pragmatic use of something that appeared to work without the doctors understanding why. However, at least by then there had been great strides made in understanding how the body functioned at the level of the organs. As restrictions preventing dissection were relaxed, the anatomy of human beings was finally studied beginning around the sixteenth century, and knowledge was gradually added to the picture of the basic mechanisms of the body. The biggest breakthrough that had happened by the 1950s in understanding disease, though, did not involve anatomy. It was the germ theory.

Microorganisms had been observed with one of the earliest microscopes back in the 1600s, but diseases that were spread through the air had been blamed on miasmas, foul vapors of rotting substances that were thought to carry the disease as part of the air. (This was the origin of the idea that "fresh air," i.e., air without a miasma, is good for you.) It had, after all, been obvious for a long time that diseases weren't all caused by what we ate or how we behaved but were spread from person to person. It's notable, for example, that the Bible contains advice on both washing the hands and keeping away from infected people in order to avoid catching disease. And vaccination had started to take off with Jenner's work using cowpox to prevent smallpox in the late eighteenth century.

During the nineteenth century, a series of discoveries brought the miasma theory into disrepute. A perfect example was the work of Dr. John Snow in London. Sometimes labeled the Medical Detective, Snow was baffled by the cholera epidemic that was stalking the London streets. He plotted the incidence of the disease on a map and found that outbreaks tended to bunch around the pumps used to provide drinking water. The water sources were being contaminated with foul output from the poorly maintained sewers, and this idea, that the disease could be spread through water, was to make miasmas gradually fade away as a concept.

Others, such as Pasteur with his work on killing microorganisms by boiling and Fleming with antibiotics, had by the 1950s led the charge in preventing microorganisms from taking hold.

Even so, when a medicine like aspirin was used to reduce pain and inflammation, no one knew how the benefits were gained. It is only in the last fifty years, with the gradual increase in understanding of body function at the cellular and molecular level, that medicine can truly be considered a science, and one that is systematically helping us go beyond the basics of Human 1.0.

The picture now is that spiritual medicine has all but disappeared from the central medical stage (though it still exists at the periphery), while chemical medicine has made huge leaps forward thanks to our increased understanding. Surgery, though vastly safer than it once was, has come less far. We don't let barbers operate anymore, but in essence, much basic surgery is still a subtle form of butchery. Most of the advances (and hence the increase in safety) have been in the introduction of anesthetics, making it possible to operate without causing huge pain and stress, in our ability to better view the body inside and out with X-rays and scans, and in the quality and precision of our instruments.

The understanding of how the body works has also enabled one surgical technique to be used successfully in a way that would never have been practical otherwise, however good the equipment—that's transplant surgery. Although work had been done with skin transplants and reattaching severed limbs earlier, the essential extra for organ transplantation was the understanding of the mechanism by which the body rejected foreign material and the development between the 1950s and 1970s of immunosuppressive drugs that reduced the body's tendency to destroy tissue that didn't "belong." The first successful heart transplant took place in 1967, in an operation led by South African surgeon Christiaan Barnard. The recipient lived for eighteen days.

Truly successful transplant surgery, with indefinite survival, has been dependent on a better understanding of dealing with rejection problems at a molecular level.

It's also true that there are a few genuinely new surgical approaches that have been devised in the twentieth century. A good example is laser eye surgery. The laser was, for a good number of years, a solution in search of a problem. In 1917 Einstein had predicted that it should be possible to set up a sort of chain reaction of light that he called stimulated emission. He imagined an atom absorbing a photon of light. When it does, an electron in the atom jumps into a higher energy state (this is the real meaning of the term "quantum leap"). At some point in time, the electron will drop back into its normal state, giving off a photon. But imagine, Einstein said, that a second photon impacts the same atom before the first photon has left. Then not only will a photon be emitted corresponding to the new photon, but the original one will be pushed out, too.

With the possible exception of his time experiment, Einstein was no practical scientist. His idea was first tried out in the laboratory in 1954 by Russian scientists Nikolai Basov and Aleksandr Prochorov. They were sending microwaves (a low-frequency form of light) through the gas ammonia in a sealed chamber and found that as predicted there was a chain effect, with more and more of the photons pulled into line by stimulated emission. They called their device a maser—Microwave Amplification by Stimulated Emission of Radiation.

The idea of an equivalent for visible light was dreamed up by two Americans independently—Arthur Schawlow and Gordon Gould, though it was Gould who was eventually recognized as the originator of this visible maser, which he named a laser. The first prototype was built by another American, Theodore Maiman, in 1960, substituting a ruby crystal for the ammonia gas used in the microwave version. Here mirrors were placed at both ends of the crystal, letting the stimulated emission build and build as the light slammed from one end of the crystal to the other, before it finally emerged from one end, which was only partially silvered.

The idea of the laser in the 1960s was typified by its dramatic appearance in the James Bond movie *Goldfinger* in 1964. It was a powerful, intense beam of light that could cut through metal and that clearly had the potential for use as a weapon or an engineering tool. No one at the time would have foreseen that most households would eventually contain several lasers—tiny low-power devices, used in a much more subtle way in CD and DVD players and recorders and in printers.

However, the cutting ability of the laser demonstrated to such effect in *Goldfinger* was to trigger the idea of using these quantum devices in surgery. Apart from one fundamental problem, lasers make ideal scalpels. The light "blade" can be made (in principle) as fine as you like, can cut through practically anything, and can automatically seal blood vessels and cauterize wounds. Unfortunately, unlike a scalpel, a laser's beam can't just stop at an arbitrary point in space. It carries on indefinitely until it meets an obstruction, which a powerful laser will then begin to cut through. This is quite different from the fictional Star Wars light saber, which couldn't sensibly be a laser, as it has a "blade" of fixed length. So using a laser to make an incision in the torso would have the unfortunate effect of cutting through the internal organs and eventually right through the body.

It's also true that, unlike the fake prop in the James Bond film, most lasers that are powerful enough to make a clean cut through solid material work in pulses, rather than a continuous beam, so a laser isn't necessarily the ideal replacement for a stainless-steel blade. Instead the lasers that have made laser eye surgery possible work on a totally different principle from a cutting blade. The laser used in such surgery is usually an excimer laser. Like the original maser, these don't work with visible light, but in this case they use the higher-frequency ultraviolet, produced in a gas mix of an inert element like argon or krypton with a highly reactive halogen gas like fluorine or chlorine.

All lasers work on matter by adding energy to the molecules—when a laser "burns" through a piece of metal, the energy of the photons of light adds energy to the electrons of the material, eventually inducing the chemical change that effectively burns the metal away. When an excimer laser hits the material that needs to be removed in an eye to correct faults like shortsightedness, it adds energy to the electrons that form the bonds in the molecules of the organic material, effectively disintegrating the tissue.

The most recent addition to the armory of those seeking to shore up our failing bodies is to work at the genetic level. Gene therapy, which I've already discussed in chapter 2, is just as relevant to medicine as it is to preventing aging. In fact, the first patient to receive the practical use of gene therapy, Ashanti DeSilva, and most of the other people to have been treated with attempts that have so far been made with this powerful but complex and hence risky approach were medical patients.

Gene therapy has already been used to increase strength in mice, modifying a gene that may help sufferers of amyotrophic lateral sclerosis (Lou Gehrig's disease), and also to protect against diabetes and reduce obesity, to give permanent "tans" to protect against skin cancer from ultraviolet exposure, and even to attempt to cure baldness.

As is obvious from the setbacks in early gene therapy, when children undergoing the therapy died, there are serious risks— but practically every new medical approach has initial problems. The first heart transplant patient died after eighteen days, but this didn't mean the whole concept of organ transplantation had to be abandoned. It will take time to get the bugs out of gene therapy—so we should be careful about getting overexcited by all the promises some breathless accounts make for the wonders of the technique—but this doesn't undermine the huge potential gene therapy has for fixing inbuilt problems in the human system. Any other approach to genetic defects is like putting a Band-Aid on a problem with a car, rather than changing the faulty part.

Gene therapy is also not limited to correcting disease-causing genetic failings at birth. Somatic, as opposed to germ line, therapy where material containing the modified genes is injected into a patient has already been used in experiments to reduce the impact of Alzheimer's disease, which affects 10 percent of people over sixty-five. As the population get older, the statistics get even worse, rising to around a third of people over ninety-five. Now more of us live longer, so more will suffer from Alzheimer's, with its traumatic effect on both the sufferers and their families.

Alzheimer's doesn't have an inherited genetic link—it isn't a true genetic disease like cystic fibrosis—but genes do contribute to its onset. Alzheimer's causes "tangles" in the structure of the brain, which result in neurons, the cells that are responsible for our thinking, dying off. In a trial begun in 2001, Mark Tuszynski and his colleagues from the University of California at San Diego inserted special neurons into the brain of a sixty-year-old Alzheimer's patient. These neurons had undergone genetic modification to add extra copies of a gene responsible for the production of a protein called nerve growth factor, which encourages neurons to grow.

The results, over a number of years, have not been a total cessation of Alzheimer's, but the therapy has slowed down its onset by a factor of 3. If further trials are successful and there is long-term safety in the treatment, it is expected this could become a significant bulwark against the onset of Alzheimer's. This example illustrates the power and the risk of gene therapy. It can provide help with conditions that are otherwise impossible to treat—yet the impact is so deep, no one can be quite sure what these extra genes will do to the brain. This emphasizes the need for caution at this stage.

Alzheimer's can't be cured by this therapy because it is not caused by a fault in a single gene (and because somatic treatment like this can never influence every single copy of a gene in the body). While a single gene being faulty or absent is responsible for some conditions, many are the result of the action or inaction of many genes and there are several other genetic factors known to be involved in Alzheimer's. For example, the protein CREB, described on page 192, necessary for the formation of memory, can help fight the degradation of memory that is a significant part of Alzheimer's if drugs are used to encourage CREB production.

While gene therapy certainly involves working at the molecular level, it requires less direct interaction with individual nerve cells than some attempts to repair damage to the brain and the nervous system. Practically any internal organ has subtleties that make it harder to fix than it might seem at first sight, but nothing else has the phenomenal complexity of the brain. This three-pound lump of tissue typically contains around 100 billion nerve cells (neurons). Bearing in mind that each of these can be connected to tens, hundreds, or even thousands of other cells, the total number of connections (synapses) in the brain runs to around 1 quadrillion.

That phenomenal number is dwarfed by the potential connections (bearing in mind synaptic connections grow and fade away) between neurons, which exceed the estimated number of atoms in the universe. If the dendrites, the branchlike connections that extend from the center of a neuron, forming the links to other neurons, were conventional wires, it has been estimated that the brain would contain ninety-three thousand miles of wiring. This is inevitably not a simple device to fix. Where there is damage to the brain itself, or to the spinal cord, restricting the possibilities for controlling the body and hence both acting and communicating, providing a repair implies interfacing with the brain in some way—a nontrivial task.

We have already seen (page 195) work on replacing damaged memory brain cells with a chip, but the opportunities here are both more wide-reaching and more likely to have a practical solution in a reasonable timescale. The simplest form of direct brain enhancement has been a procedure called deep brain stimulation, which has been in use in the United States since 1997 to help control the symptoms of a number of problems—initially it was developed to help with a disorder called essential tremor and more recently it has been used with Parkinson's disease. In 2007 this was extended to cover a debilitating disorder called cluster headaches.

In deep brain stimulation electrodes are implanted in one of the parts of the brain responsible for movement such as the thalamus and the globus pallidus. Before implantation, the neurosurgeon uses a brain scan—magnetic resonance imaging or computer tomography—to identify exactly where in the brain the signals are coming from that generate the unwanted tremors or pain, before inserting an electrode into the appropriate location.

The electrode is connected to a small pacemaker-like electronic device implanted under the skin, just below the collarbone. The electronics generate a regular electrical pulse, which interferes with the signals in the brain that are causing problems and has been shown to improve the motor control (or reduce the crippling headaches) of the sufferers. Most medical authorities are still cautious about this treatment—although it has now been used on thousands of people, there is some concern that in the long term the brain will get used to the stimulation and will no longer respond to it.

The same treatment is also being looked into as a possible treatment for severe depression and for obsessive-compulsive disorder. Depending on the location that the electrode is placed in, it was noticed (purely as a side effect of treatment for muscle disorders) that the patient felt either depressed or elated when the stimulation was used. As a result of this, there is some hope that the same treatment can be used to help sufferers who have otherwise untreatable depression.

Though obviously related, this isn't quite the same as experiments undertaken in the past by Robert G. Heath and others, working with animals and humans, that have shown that it is possible to stimulate pleasure or pain using electrodes inserted into the appropriate regions of the brain. Although these experiments were aimed at entertainment rather than curing illness, it is not impossible that mood-controlling stimuli will be used in the future, either legally or as a twenty-first-century equivalent of a narcotic. While at the moment this is only practical with implantation of electrodes, it would be much more likely to become widespread—almost inevitable, in fact—if inductive equivalents of EEG (see page 198), allowing this kind of stimulation without brain operations, were developed.

More sophisticated brain enhancements don't try to regulate the existing parts of the brain but go further, adding on functionality. Unlike the simple upgrading provided by an early tool like a walking stick, this kind of enhancement involves more than supporting an inbuilt function—it bypasses the inbuilt mechanism and provides a whole new way of fulfilling a need, a change that in evolutionary terms is particularly sophisticated.

This kind of enhancement, making it possible to link an artificial extension to the brain, is known as a neuroprosthetic. These aren't as new as they might seem—the first neuroprosthetics were introduced in the 1960s, to overcome some types of deafness. When we hear a sound, the complex pressure waves in the air are translated into sets of firing neurons by thousands of tiny hair cells in the cochlea. This is a spiral-shaped bone chamber (the word comes from the Latin for "snail"), which is filled with a watery fluid.

Sounds coming into the ear vibrate the eardrum, which passes the movement through three tiny bones (the smallest in the body) onto a membrane called the oval window, which sets the fluid in the cochlea in motion. This movement is picked up by the tiny tufts, which look a little like hairs but are actually extensions of cell membranes that stimulate the "hair cells" at their base and generate signals in the auditory nerve. If these hair cells are damaged, hearing loss occurs.

Cochlear implants get around the failure by directly stimulating the neurons that the hair cells should act on. An external headset picks up sounds and processes the signal to produce a series of electrical impulses, which are linked by induction to a small device implanted under the skin that stimulates electrodes embedded in the cochlea. The earliest implants only had a single electrode, though numbers have increased over time to twenty-plus separate stimulation points. Even so, only a small subset of the hair cell–connected nerves are activated, and it was originally assumed that the implants would just give a little guidance to help with lipreading, but in fact they have enabled users to understand speech, proving much more effective than was originally expected. Over one hundred thousand people have now benefited from cochlear implants.

A second sensory area in which practical experimentation on human beings has already been undertaken is sight. The self-funded researcher William Dobelle made his first brain implant in 1978, giving a man named Jerry a limited approximation to sight using TV cameras mounted on the frame of his glasses, generating signals that were sent via a mainframe computer to electrodes implanted in his visual cortex. Over the years, Dobelle has refined his technology to a more portable system.

There is some controversy over Dobelle's work, as he works outside the academic framework and charges big money for the treatment. Working alone, he lacks the constructive synergy that groups in universities have through publication and teamwork. Even so, whether or not Dobelle's specific devices will ever become commonplace in the medical world, he has proved that an equivalent to a cochlear implant for eyesight is not impossible, though the need to link into the brain proper makes this a more invasive technology. His system doesn't restore sight in the normal sense but rather gives an approximation to sight that is sufficiently good for basic awareness of the surroundings.

When the electrodes in Dobelle's device stimulate neurons in the visual cortex, the result is the production of phosphenes, small bright areas of apparent vision. We've all experienced phosphenes when we rub our eyes and little flashes of light appear to float in space before us. Here the optical nerve is being stimulated by pressure. With Dobelle's implants, electrical impulses provide the stimulation. The initial calibration of his system involves putting a series of signals through the electrodes. The subject describes what he or she "sees," and the researcher builds up a map of these illuminations, rather like the pixels on a computer screen, except the phosphenes are much less regular in shape.

Once a complete map has been built, Dobelle's software can make use of the phosphene "pixels" to display images from the video camera that the subject wears. In the most recent version of his technology, there are two cameras, accessing two sets of electrodes on the left and right side of the brain, just as our optic nerves cross over to the opposite sides of the brain. Even with this latest version, the effect is limited, and at least one subject has had a seizure as a result of excessive stimulus to the brain, yet it does make it possible for the wearer to identify objects and to manipulate them.

Dobelle is not alone in this work—a number of universities have projects under way, and there is no doubt that there will be visual implants available much more widely within ten or twenty years. Some of the experiments go even deeper into in-brain electrodes. Others are looking at stimulating the optic nerve externally, holding out the hope for some blind people (depending on the cause of the blindness) to have restoration of elements of sight without the need for intrusive brain surgery, just as cochlear implants can restore hearing without putting electrodes into the brain itself.

Once it is possible to feed the senses electronically, particularly if it is possible to do so without implanting electrodes, there is no need to stop at the limits of normal human capabilities. We see a tiny fraction of the spectrum of light, our eyes only reacting to wavelengths between around four and seven hundred nanometers (a nanometer is a billionth of a meter). Extending below with longer wavelengths are infrared, microwaves, and radio. Above come ultraviolet, X-rays, and gamma rays. With the right camera, the same neuroprosthetic that enables basic sight could see the heat pattern of people in the dark or provide the equivalent of using night-vision goggles but linking directly to the brain.

Similarly, our hearing cuts off at a relatively low frequency—with an appropriate microphone, a cochlear implant could pick up the echo-locating squeaks of bats. As well as being able to see in the dark, these technologies, coupled with appropriate sensors, would even enable us to stray beyond traditional sensory boundaries. It has been suggested, for instance, that a dog's sense of smell is more tightly linked to its eyesight than ours, giving it an almost visual sense of smell that allows the animal to navigate by odor. Although it's natural to link light input to the eyes, we could equally "see" sounds from an electronic ear or smells picked up by an electronic nose—it is entirely possible that a future sensory neuroprosthetic could give the wearer a much richer interaction with his or her environment than we are naturally capable of with normal human senses alone.

Similarly, the assumption tends to be that the pickups for the senses will be located in the natural place for them to reside—on the user's head. But with modern electronic communications there is no need for this to be the case. The sensory inputs could be coming from a tank on a battlefield the other side of the globe—or from a buggy on the surface of Mars. (The only trouble here is the time delay. Light, our fastest means of sending a signal, takes an average of four minutes to reach us from Mars, and another four minutes is taken up in getting a signal back, so the driver would need plenty of time to react to a problem.) And that's only the start. The signals entering the brain needn't even be linked to a real sensory input at all.

Where watching a movie is always a more restricted experience than seeing the real world, the users of the brain-linked equivalent of a movie would see no difference between an electronic source and true vision. What's more, the image "seen" need not ever have existed in reality. It could be the virtual world of a computer game or the unreal universe that is formed by the World Wide Web. If the quality of images can be brought anywhere near that experienced in true vision, and particularly if it becomes possible to interface with the brain without surgery and electrodes (see page 230), this technology could transform how we interact with both the real and the virtual world.

In 2006 two papers published in the journal *Nature* gave hope for a much wider use of neuroprosthetics to help those with damage to the nervous system regain the ability to interact with their environment. In the first, John Donoghue and colleagues at Brown University described how they implanted an array of ninety-six electrodes into the precentral gyrus area of the primary motor cortex, the part of the brain responsible for movement. The experimental subject, Matt Nagle, was a man whose spinal cord had been entirely severed in an accident, leaving him with no control over his limbs.

Even though the experiment took place three years after Nagle's injury, he was able to "think" hand movement and produce signals through the electrodes that were able to move a cursor on a screen, control simulated e-mail, and operate connected devices such as a TV. He was able to do this while talking, as naturally as someone without his injury could talk and move his hand at the same time. Nagle was also able to open and close a prosthetic hand and make basic movements with a robotic arm.

It might be imagined that this took a huge amount of training, gradually working up the tiniest elements of control to eventually achieve something usable (it's certainly how Hollywood would show it, probably in a montage with dramatic music in the background), but in fact one of the surprises for the researchers was just how quickly Nagle gained control. Within minutes of being asked to imagine using his hand to move a cursor on the screen, he had succeeded, and then, without further training, he was immediately able to move on to perform the other tasks mentioned with a degree of success.

This wasn't the first ever attempt at direct control from the brain—an experiment in 2000 used just two electrodes to provide very simple control of a cursor—but the work with Nagle was remarkably effective for such an early stage of the process of developing physical neuroprosthetics, and it holds out great hopes for true "bionic" limbs, controlled by thought. It also makes feasible a much wider extension of thought control beyond replacing existing functions to enable use of many different devices without physical interaction.

Although the experiment involving Matt Nagle took place at Brown University, it was undertaken in cooperation with the biotechnology company Cyberkinetics and the equipment used already has a trade name, BrainGate, demonstrating the commercial plans of those involved. Even so, it clearly would be desirable to use a noninvasive connection. We have no experience to suggest how such fine electrodes would stand up to lifetime use, and there is no doubt that there is an accompanying risk of infection and from the insertion operation that makes it highly desirable to work toward an external wearable connection. What's more, there is some evidence that for unknown reasons the response recorded by direct electrodes drops off after a time, and different individuals seem to have different capabilities in using them (though this may just reflect subtly different placement of electrodes).

The second paper in this field featured in *Nature* in 2006 demonstrates the ability to go beyond the explicit concept of mentally moving a hand to the intentions behind it. Krishna Shenoy's group at Stanford University used a new approach with a series of electrodes in monkey brains to capture the intention of motion much more precisely. By using signals from the premotor cortex, they were able to predict the position of spatial targets the monkey was aiming for. This meant a significant speeding up of the link from brain to action. The researchers suggest the accuracy and speed provided mean that it would be possible to mentally control typing at a speed of around fifteen words per minute if such a system were available to a human brain.

The significance of this development is that where someone using the BrainGate approach would have to consciously control an invisible hand—fine to move a cursor but very slow for an action like typing—this approach would make it possible to think about typing and to have the appropriate signals produced, rather than consciously moving a virtual finger from key to key.

These experiments show the first steps toward being able to directly control robotic limbs from the brain—but to truly replace a physical limb there also needs to be appropriate feedback. We don't just need to be able to move a limb; it's also necessary to know where it is, an ability that's not as obvious as it seems. This requirement is important both for direct limb replacement and also for remote control of a robot or to get feedback from the sort of exoskeleton DARPA envisages being used by the military of the future (see page 157).

We have an awareness of the location of the parts of our body that is essential for fine control. It's this feedback that enables us to touch our noses with our eyes closed or, more practically, to direct our body to take action without consciously watching exactly what a finger or leg is doing. Although no work has been published yet, there is research going on in a few labs to try to close the loop and supply feedback to the brain, providing the kind of information that the nervous system channels from an arm or a leg.

The success of cochlear implants holds out some hope that scientists won't have to be too precise about matching the exact inputs provided normally by the nervous system. The brain seems capable of handling highly limited signals and sorting out the necessary interpretation, though of course the hope is to match "real world" inputs as closely as possible. At this stage it's not even clear where the signals would be inserted into the nervous system—these are very early days. Experiments are being undertaken with animals both by moving limbs and recording the response in the brain, which will then be mimicked, and by stimulating parts of the brain that correspond to feedback from the limbs and seeing how the motor cortex responds.

It will take time, but from existing developments it does seem entirely possible that within ten to twenty years we should be able to have the first naturally controlled artificial limbs or a remote robot controlled by mind with feedback that gives the feel of actually being that device. It shows just how far ahead of reality fiction like *The Six Million Dollar Man* and practically every cyborg you see on a sci-fi show is—but equally these developments emphasize that these concepts are no longer pure fantasy. In our quest to improve on basic *Homo sapiens* it seems that almost anything is possible in the next century or so.

Although every step along the way toward production of a cyborg seems positive—overcoming injuries, treating disability, adding strength and better sensory capabilities—the reaction of almost everyone to a fictional cyborg is one of horror, disgust, and sometimes pathos. In theory a cyborg can be more than human, but unless the electronic and mechanical implants are all subcutaneous (as in *The Six Million Dollar Man*), the result is the production of a visual monstrosity. There is something unnerving about the Borg in *Star Trek,* for example, an insectlike mechanical disfigurement of the regular human form.

Cyborgs aren't the only monsters that those who dislike the concept of human enhancement worry about. There are very many ways, they believe, that we can take humanity too far and produce a creature as shocking as anything dreamed up by Mary Shelley in *Frankenstein*.

7.

Monsters and Mutants

I beheld the wretch—the miserable monster whom I had created.

—Mary Shelley, *Frankenstein*

The five components we explored in the previous chapters, each making real our need to change ourselves in some way, have pushed ahead the ceaseless development of *Homo sapiens*. Together they combine to make us much more than the biological milepost fixed by evolution one hundred thousand years in the past. But is the unnatural quality of this driven change putting us in danger? Are we becoming monsters?

Biologically speaking, every one of us is a particular kind of monster—a mutant. We all carry tiny errors in our genetic codes. Mostly these go unnoticed, but occasionally a mutation will produce something new, or something horrible. In nature, most monsters prove to be dead ends. The majority of drastic mutations either are incapable of living or so reduce the functional effectiveness of the individual that the new genes will not be carried forward far into the gene pool. In history, human mutants were often called monsters or monstrous, their unfortunate deformity considered the work of the devil, or at the very least providing a specious omen of ill times.

Some of the descriptions of human mutants from earlier times show the exaggerative imagination used in nineteenth- and twentieth-century freak shows, a delight in monstrosity that was all too common before a changed view of what was acceptable banished these terrible exhibitions from the carnival midway. When a poor child was born with some form of deformity in medieval times, rumor would quickly transform the child to something altogether more dramatic. In 1540, for example, a baby born in Cracow (now Krakow, Poland) would become known as the "monster of Cracow." It was said that the baby had barking dogs' heads on its elbows, chest, and knees. Clearly this was fantasy, but the baby probably suffered from some form of disfiguring mutation.

Our self-enhancement is different. Because it has conscious direction, there is much less of a tendency to produce impractical monstrosity. You will never see an enhanced human being looking like a 1950s sci-fi alien with a bulging exposed brain and a tiny withered body—not only does this move run counter to the second evolutionary driver of attractiveness, but also any such development would be naturally unsustainable. The mutant just wouldn't survive.

Mutations are changes to our genes, usually caused by a fault in the genetic copying mechanism that results in an imperfect set of information in a copy of our DNA, or caused by an attempted fix of faulty DNA that fails. Some such changes will have a major impact across many genes, perhaps even resulting in an extra chromosome. Others will operate at the level of a single letter of the genetic code. Some human mutation is entirely harmless. I have red hair—this is not a "natural" human hair color but a mutation that is common in Celtic races. Apart from a little bullying at primary school, it hasn't done me any harm.

Other mutations can result in life-threatening diseases such as cystic fibrosis, mental problems, or deformities that make it impractical for a human being to live. Many mutations switch genes off, stopping a process or the development of an organ that should be part of the human makeup from ever existing. Others add something, perhaps producing a protein that wouldn't otherwise have been made with results that can be useful or dangerous. A mutation might, for instance, give a natural resistance to a disease. When this happens, if the disease is a life-threatening one, the mutation will tend to spread through the population by natural selection, as more of those with the mutation will survive. This is happening at the moment in Africa with a mutation of the CCR5 gene that results in improved resistance to HIV/AIDS.

On average, each child has around one hundred mutations that weren't passed on to him or her from the parents. Most won't do anything—they will effectively be "junk" DNA, a section of the genetic code (and there are many of them) that has no active part to play in the development of a human being. Around three will have a negative effect, causing a change to the way that genes specify proteins that will be a disadvantage, while typically one of them will have a positive effect. Mutation is a biological lottery that we all take part in, whether we like it or not.

The most immediate threat to human beings from upgrading comes not from a direct modification of the human form but through external enhancements, the technology that we have built around ourselves to extend our capabilities. Pretty well every piece of technology brings dangers as well as benefits. Two of the technologies we looked at earlier were the dog and the flying machine. Most of the time, dogs are very safe technology, but every now and then one will "go wrong" and bite a human or run out into a road, causing a car crash.

Similarly, flight technology is now very safe. The statisticians keep telling us that it's the safest way to travel. This is, incidentally, an example of spin from the airline industry, because it's certainly not the only interpretation that you can give the statistics. It's useful spin because we are more aware of plane crashes than car crashes. When a plane comes down, there are often hundreds of people killed, but car crashes are everyday occurrences. This makes plane crashes seem more significant than they really are, playing down the fact that around forty thousand people are year are killed on the road in the United States alone, compared with a much smaller figure killed in planes.

The reason the statistics can be misleading is that they are based on the statement that you are more likely to be killed in a car crash in any particular year than in a plane accident. But what seems more important to an individual is "am I likely to be killed on the flight I'm about to take?" or "will I survive this car journey?" If you make that comparison, trip by trip, driving is a little safer—we make so many more car journeys than plane flights that any particular car journey is a little safer than a particular flight. Even so, the risk of being killed on a flight is only about one in 8 million, so it shouldn't keep anyone awake at nights.

However, though we shouldn't be worrying because we've a plane trip coming up, there is still a risk that it will hurt or kill us—and the technology of flight is also used to deliver a host of deadly devices from the air, whether as bombs from a plane or in the form of a missile—so there is little doubt that flight can cause us harm as well as good.

Perhaps the most drastic risk to date from our technological enhancements is one we face from the use of nuclear power, whether employed in power stations that can give us Three Mile Island or Chernobyl or in the stark threat of the nuclear weapon. And the danger we face at the hands of our technology extends, perhaps just as threateningly, to one of the new technologies that is the heart of some of the human modification foreseen by the likes of Ray Kurzweil—nanotechnology.

Imagine an army of self-replicating robots, each invisibly small, endlessly reproducing, forming a gray mass that swamps the world and destroys its resources. Each tiny robot eats up natural resources in competition with living things and reproduces at an inhumanly fast rate. This is the premise of Michael Crichton's thriller *Prey,* but it isn't an idea that is limited to science fiction. Nanotechnology has a huge potential for good—but can also cause horrendous damage as this so-called "gray goo" scenario shows (gray goo because the nanobots are too small to be seen individually and would collectively appear as a viscous, self-moving gray liquid).

Ever since Antoni van Leeuwenhoek peered through his early microscope in 1674 and saw "animacules"—microorganisms—the world of the very small has fascinated humanity. The concept that there could be creatures and constructs so small that we could not see them was boosted into a central theme of physics as atomic theory came to the fore. Initially, manipulating incredibly small objects directly seemed impossible. Work on atoms and molecules could only be undertaken on the mass scale of accelerators and atom smashers—early speculation assumed that we ourselves would have to be shrunk in size if such interaction were ever to be possible. So in Isaac Asimov's *Fantastic Voyage,* for example, we saw miniaturized humans interacting with the components of the human body.

Until recently, the prefix "nano" was an unfamiliar one. At the 11th Conférence Générale des Poids et Mesures in 1960 a faceless committee defined the SI (Système International) units. As well as agreeing on standards of measurement such as the meter, the kilogram, and the second, the conference developed a range of prefixes for bigger and smaller units from "tera" (multiply by 1,000,000,000,000) to "pico" (divide by 1,000,000,000,000). The penultimate prefix was "nano" (divide by 1,000,000,000), derived from *nanos,* the Greek word for "dwarf." One-billionth, a truly tiny scale.

Twenty-six years later, American author K. Eric Drexler combined the "nano" prefix with "technology" in his book *Engines of Creation.* Although Drexler would outline a wide range of possibilities for products smaller than a microbe, the majority of the book and the publicity that accompanied it focused on molecular manufacturing, using nanomachines to assemble objects at the molecular level, a concept first suggested by physicist Richard Feynman. A single assembler working at this scale would take thousands of years to achieve anything—effective assembly would require trillions of nanomachines. Drexler speculated that this would result in nanomachines that could replicate like a biological creature, leading to the vision of gray goo and Crichton's *Prey.*

Nanotechnology itself doesn't necessarily involve anything so complex as an assembler. One very limited form of nanotechnology is already widely used—nanoparticles. Substances reduced to particles on this scale behave very differently from normal materials. The most common use of nanotechnology currently is sunscreens, where nano-sized particles of zinc oxide or titanium dioxide are used to protect us from the sun's rays, allowing visible light to pass through but blocking harmful ultraviolet. In fact, we have used nanoparticles unwittingly for centuries in some of the pigments used in pottery glazing.

Now on the way, with vast potential, are nanotubes and fibers. Often made of carbon, these molecular filaments are grown rather than constructed and have the capability to provide both superstrong materials (as an extension of the current cruder carbon fibers) and incredibly thin conductors for future generations of electronics. Semiconducting nanotubes have already been built into (otherwise) impossibly small transistors, while carbon nanotubes could make one of the more remarkable speculations of science fiction a reality.

Writer Arthur C. Clarke conceived of a space elevator, a one-hundred-thousand-kilometer cable stretching into space that could haul satellites and spacecraft out beyond the Earth's gravity without the need for expensive and dangerous rocketry. Bradley Edwards, working for the NASA Institute for Advanced Concepts, commented in 2002: "[With nanotubes] I'm convinced that the space elevator is practical and doable. In 12 years, we could be launching tons of payload every three days, at just a little over a couple hundred dollars a pound."

Yet the ultimate aim of the nanotechnologist is the nanobot. These are Kurzweil's tiny machines that would be injected into the bloodstream to fix our cells, or tiny assemblers that take material and reassemble the atoms. In principle an assembler could make anything, unstitching atom from atom and reconstructing the building blocks of nature into anything from a scarf, to a TV, to a piece of beef. Each individual assembler could only handle a tiny fraction of a whole object, so to make something we can use we need billions of nanobot assemblers.

If we are to achieve Kurzweil's dream of tiny intelligent machines inside our body that keep us alive, we probably have to go some way down the assembler route. The only sensible way to build such nanobots is to use other nanomachines. Yet this doesn't mean for certain that we will plunge into a nightmare future where these self-replicating devices take over the world. If they can ever be built (and there's always a possibility that, like the walnut-sized personal nuclear reactors imagined in the 1950s by science fiction writer Isaac Asimov, they simply aren't practical), these tiny constructs will sometimes cause damage, yet there is no reason that they can't be built with just as many safeguards as any other piece of technology.

The biggest issue comes from the replication, because we have a clear model for how things can go wrong in the natural world, something we already met at the start of this chapter—mutation. When something is being replicated many, many times, there is a chance of an error in copying. Usually that error will result in failure, but occasionally it can make a change that will make the replicating creature better—natural selection will ensure that the "better" form of the creature, assuming it can pass on its difference, thrives and takes over. That's evolution in a nutshell. It happens with the biological machines that populate the world, and it could happen to nanomachines. Once machines have the ability to replicate, and to pass on changes in design, they can evolve.

The "gray goo" scenario envisages a nanomachine that has,

thanks to a random error, gotten out of control, producing a "better" machine from its own point if view, though not for humanity, as it simply duplicates itself and consumes, becoming a ravening horde that destroys everything from crops to human flesh. It's a ghoulish thought. Yet the parallel with biology isn't exact. The big difference between a machine and a plant or animal is that the machine is designed. And the right design can add in many layers of safeguard.

First there is the resistance to error. Biological "devices" are much more prone to error than electronic ones. Yes, there could still be a copying error in producing new nanobots, but it will happen much less often.

Second comes error checking. Our biological mechanisms do have some error checking, but it doesn't stop mutation. It is entirely possible to build error checks into electronic devices that prevent a copy from being activated if there was an error in the copying. Depending on the level of risk, we can implement as many error checks as we like. This is a crucial difference between design—where we can anticipate a requirement and build something in—and the blind progress of evolution.

Third, we can restrict the number of times a device can duplicate itself, as a fallback against rampant gray goo. Finally, we can equip devices with as many other failsafes as we like. For instance, it would be possible for nanomachines to have built-in deactivators controlled by a radio signal. All these layers of precautionary design would make any nanobots much less of a threat than those who find them frightening would suggest. Perhaps the biggest danger is not from accidental evolution of a nanobot but from the intentional handiwork of a hacker.

Every day in my e-mail in-box, in addition to the genuine

e-mails I want to receive I get a whole host of others I don't want: spam trying to sell me Viagra or fake watches and, worse still, e-mails carrying viruses, trojans, and worms, all hoping to take over my PC or cause damage. Some virus writers produce them for fun or as an intellectual challenge, but other virus writers are, in effect, electronic terrorists who hope to cause disruption and confusion. The result of this relentless impact from would-be attackers is that I have to have three programs running all the time—antivirus, a firewall, and antispyware, constantly battling to protect my computer.

These electronic vandals don't stick to a single technology. As long as it's widespread enough to be worth their attention (the reason Apple computers are relatively unscathed), they will get involved. Now that the more expensive cell phones are in effect pocket computers, virus writers have turned their attentions to your pocket. In June 2004 the first cell-phone virus emerged into the wild. One of the particularly unnerving things about cell-phone viruses is that they are more like the real thing than anything that arrives on your computer.

The use of the term "virus" when referring to a malicious computer program has always caused confusion. When computer viruses were first made public I ran the PC department of a large company and once had a phone call from a worried executive who had recently become pregnant and was worried about catching the computer virus and the danger of it causing damage to her unborn child. Cell-phone viruses don't put humans at direct risk any more than computer viruses, but they certainly can hit our wallets, and they can spread in a worryingly natural manner.

Here's a typical scenario featuring an attack by the phone worm CommWarrior. You are in a bar and need to make an urgent call. Your cell phone beeps—it asks if you want to accept a Bluetooth connection from someone you don't know. You click "No." But before you can do anything else, up comes the request again. It keeps coming so fast that you can't place your call. So you finally say, "Yes," to get it out of the way—and you're infected. CommWarrior has jumped from a stranger's phone to yours. Because of the way Bluetooth works, cell-phone viruses and worms that use it jump from person to person when they are in close proximity. You literally catch a cell-phone bug like this by being near someone who is infected (though that person doesn't have to breathe on you). And once you are infected, the virus can start siphoning cash from your account.

It would be naive to think that hackers who can jump on the bandwagon so effectively with smart phones won't do the same with nanomachines. If anything is going to give us cause to pause and think whether or not we want to go down this route, it is the possibility of hacking. Yet even this can be avoided. Computer viruses, whether on a PC or a phone, are just programs. It's entirely possible to make intelligent electronic devices that can't have a program run on them other than the one that is built in. We normally allow reprogramming because software often needs updating and we don't want to throw the hardware away if the program is wrong—but nanobots are designed according to a different model. They don't need to be reprogrammed—we can literally rebuild them molecule by molecule instead. This both is an antihacking advantage and makes the software less complicated.

With nonprogrammable nanobots we can avoid hacking. What isn't possible, though, is to prevent malicious people from producing their own nanomachines that can cause damage—but the genie is already out of the bottle. Just as it was impossible to forget the possibility of atomic weapons once the concept had been devised, we can't go back to a time when the idea of nanobots hadn't occurred to anyone. We are already at the stage of "the good guys had better build them, or it will only be the bad guys doing it."

It's also worth stressing again that we may never be able to build effective nanomachines. There have been experiments producing promising components—for example, nanogears assembled out of molecules, and nanoshears, special molecules like a pair of scissors that can be used to modify other molecules—but just think of how much further we have to go. Not only do we need to build something complex on this scale, but we also need to give it a power source, a computer, and the mechanism to reproduce. Of these we can only manage to build the computer at normal macro scales at the moment, let alone something so small we can't see. (Yes, we have power sources, but battery technology is clumsy and big and runs out very quickly.)

Even if the technology were here today, there would probably be scaling issues. Just as you can't blow a spider up to human size because of scaling, you can't shrink something that works at the human scale down to the nano level and expect it to function the same way. Different physical effects come into play. The electromagnetic influences of the positive and negative charges on parts of atoms begin to have a noticeable impact. A quantum process called the Casimir effect means that conductors that are very close to one another on the nanoscale become powerfully attracted to one another. At this very small scale, things stick together when they shouldn't. This could easily mess up nanomachines, unless like their biological equivalents they resort to greater use of fluids. Almost everyone making excited predictions about the impressive capabilities of nanomachines wildly underestimates the complexity of operating at this scale.

Nanotechnology may seem frightening, but some futurologists envisage a more terrifying world where the inevitable conclusion of our technological developments will be a total replacement of humanity. Way back in 1995, a trio working at the new technology labs of the British telecommunication company BT came up with a worrying prediction for the "future evolution of man." Ian Pearson, Chris Winter, and Peter Cochrane imagined that by around 2015 computing and robotics would be so advanced that *Homo sapiens* would be overtaken in all abilities by a robot species they dubbed *Robotus primus*.

At the same time they imagined human beings becoming sufficiently enhanced with electronic implants to be considered a new species themselves—*Homo cyberneticus*. But this first cyborg they felt was a transitional move to *Homo hybridus*, which would have biological enhancements as well as the cybernetic (back in 1995 it wasn't obvious that genetic engineering would race ahead of the interface between man and machine). This hybrid, too, was expected to be relatively short-lived. As the capabilities of electronics continued to grow, the parts of the hybrid where normal biological processes were better than the electronic would become fewer and fewer. Between 2100 and 2150, Pearson, Winter, and Cochrane imagined, we would abandon flesh and become purely electronic—*Homo machinus*.

This new species would be quite different from anything we know. In the BT trio's words, the citizens of the future would see a species that "is vastly more intelligent and has access to whatever physical capability is required. It can travel at the speed of light, exist in many places at once, and would be essentially immortal. It would coexist with Robotus primus, but we could expect that the two would closely interact and may quickly converge."

They also envisaged that many humans would reject enhancement, living in parallel with the increasingly enhanced forms, though they predicted that peaceful coexistence would be unlikely for too long and the remaining *Homo sapiens* (or *Homo ludditus,* as they christened them), would be rendered extinct by their creations before long.

If this picture—a merger between the Terminator and Matrix movies and Star Trek's Borg—seems a frightening one, it needs to be tempered with a few provisos. The timescales now seem wildly optimistic. We are more than halfway through their route to the development of a robot that can outdo us, a development that seems unlikely in that timescale (see page 288). It's also worth reminding ourselves that one thing futurologists have in common is they always seem to get something big wrong. It is possible that this view of the future could be correct, but if it is, living it will be very different from the prediction—what may seem monstrous to us now would be anything but for those experiencing it.

We don't have to look to such extreme examples, or to our external enhancements at all, for every danger that we face through modification, though. As we move to make more direct changes to ourselves, there are real dangers posed by genetic engineering. Researchers at the Salk Institute at La Jolla, California, have shown that the three children already mentioned who developed leukemia after gene therapy for a rare immune disorder probably developed the cancer because of an interaction between the inserted gene and a nearby existing gene. The human body is such a complex mechanism that it is easy to be unaware of all the consequences of an action. But the same goes for a drug trial or any other attempt to modify our bodies. Any individual action may cause damage, distress, or even death to the individual—though it's less of a concern for a species.

A more realistic issue for our species is raised by bioethicists such as James Hughes of Trinity College in Hartford, Connecticut. They fear that we may end up with an underclass, with a division between the modified and the unmodified. George Annas of Boston University takes this even further, asking if an enhanced subset of the population will "see other people as subhuman, and enslave or slaughter them?" There are two faults in this analysis when looking to the future. One is that Hughes and Annas seem to miss the fact that everyone has already been upgraded. The other is that there are already plenty of divisions we can make in human beings if we want to. In history, we have been all too quick to take this approach to other races, even though they were from the same species, but now we are significantly better at overcoming this tendency.

The Nazis infamously tried to produce an "Aryan" state, treating (just as Annas comments) others as subhuman, enslaving and slaughtering them. Yet this isn't an argument for not having human diversity—it is an argument for practicing tolerance. We are much better at this (at least in a fair percentage of the world) than was the case in the middle of the twentieth century when there was still segregation in parts of the United States. It seems a backward-looking step to suggest that just because some future humans will be different they will immediately revert to the behavior of an earlier age. Of course there will always be gaps between those who are at the leading edge and the masses, but the furious development of technology has seen an equally quick growth in the rate at which improvements become widely available.

The most obvious split between the "haves" and the "have nots" is between the West and the Third World. Looking at the specific component of life expectancy, which it is easier to make a comparison with than beauty or strength or brain power, the entire world had a similar expectation of life span up until the eighteenth century. Average length of life might have dipped as low as eighteen and scratched its way up to twenty-eight, but on the whole it meandered around the twenty-five-year mark. Twenty-five years—a length of time that seems pitifully small today, particularly if, like me, you have already passed the fifty mark.

As we have seen, average life expectancy is a misleading measure. The very low average reflects the fact that many, many people died in childhood. For example, in England in 1750, around two-thirds of children did not survive to age five. If they survived to adulthood, there were still many more chances of dying from illness and accident through the middle years, but should a U.S. male last to seventy back in 1900, he would be expected to live to seventy-nine, where now he could hope to reach eighty-two.

Even so, the increase in average life expectancy does represent one way that we have escaped from our "natural" state, going beyond the evolutionary picture thanks to the application of our ingenuity. From the mid-1700s, the life expectancy in the West began to diverge from that in less-developed countries. The steady increase in the West was not initially paralleled in developing countries. There was a lag of around 150 years before the figures everywhere began to climb. Since then, though, both have risen roughly in parallel, with the life expectancy in the developing world rising significantly faster in the last century than that in the developed world has.

Although it is arguably unfair that life expectancy can still vary so much in different parts of the world, this unfairness is not an argument for refusing to take on the advantages of further life extension, as these benefits, too, can be expected to spread around the world, and based on the experience with general life expectancy, the gap between the developed and developing world may well narrow further still over time.

Initially, any human enhancements will tend to be most common in developed countries. This isn't purely a matter of concentration of wealth. Even where the enhancement has little or no cost, to become commonly used there needs to be a spread of information and an acceptance that this is something worth spending time and effort on. If you are near the poverty line, the basics of where the next meal will come from tend to overpower the importance of eating the right balanced diet. Similarly, when it comes to accepting human upgrades, individuals need to be focused beyond surviving the next twenty-four hours (even though those same upgrades may help with that time span). Over time, though, costs fall, information spreads, and the upgrade becomes more and more universal.

Take the example of flying. Initially this was a field purely for technical experts. With commercialization it spread to the wealthy who could afford their own plane or could pay the exorbitant prices of the early aviation business. By the 1970s, many in the West could afford a flight or two a year, though it was still a province of the rich to use airplanes like buses. Now deregulation and economies of scale have brought cheap flights to an ever-extending market—and this technology is spreading rapidly to the developing world. India and China already have their first budget airlines, and these huge countries will rapidly become in-depth consumers of this dramatic human upgrade (climate change permitting).

Certainly there will be some divide between rich and poor, but this is not a distinction that derives from our ability to enhance ourselves but rather to our allegiance to a capitalist society. Capitalism and free markets lead to a degree of unfairness, it's true. Yet no one has ever come up with a workable alternative that is anywhere near as good. Unless we are prepared to say that there should be no financial inheritance—that rich people are not allowed to give anything to their children—the children of the better off will always have some form of advantage, a starting point in life that gives them better chances. This doesn't stop the person from a poor background who has real drive and determination from succeeding, but few would argue that a billionaire's children are likely to lack a certain advantage in life.

This same unfairness does come through in genetic modification—it will be those with plenty of money who will get it first (though at the speed with which costs in this arena fall, what costs a million dollars one year might cost just a dollar a few years down the line), but it's not sensible to blame that inequality on genetic enhancement itself or to ban the process simply because it will be available first to those with money, any more than we should ban fast cars or yachts because they tend to be bought by the rich.

In the broad picture, then, there is not too much to worry about in terms of inequality. Yet there are some specifics that need to be addressed. If brain-enhancing drugs make those who take them more capable of passing examinations, are such tests still a fair way to make a selection? Should we, as in sport, have doping tests, so that entrants to university compete on a level playing field? Or should we accept that these enhancers will be part of everyday life and no more attempt to interfere than we would with, say, nutrition, arguing that someone on a healthy diet will probably do better than a candidate dedicated to junk food, so everyone should eat junk?

No one in their right mind is going to suggest we have diet tests, submitting blood samples to machines that will ring alarm bells if a candidate has eaten the right kind of things to be alert and full of energy. Yet it's arguable that the same criterion should apply to the sort of brain enhancements we can envisage coming along over the next decades.

Then there's the more dramatic creation of a true underclass if our modifications become so great that a modified human becomes to all intents and purposes a different creature. What does this do to the person being modified, and how does it affect those who are left behind? Even the simplest changes, it is argued by some, chip away at our humanity. If, for example, drugs and memory chips could make it possible to pass any examination without studying, doesn't this reduce the humanity of the individual, who no longer learns the benefits of striving? Isn't it a recipe for producing at best a human race that expects everything on a plate—and so is potentially wiped out by the next real challenge it faces—and at worst a world where there is nothing with any meaning and nothing to strive for?

Some argue this is true, but their complaint is nothing new. It was the same argument voiced by textile workers in England in the early nineteenth century when they saw the repetitive tasks that filled their days being taken over by machines. In the town of Nottingham in 1811, word spread that a man named Ned Ludd had taken action into his own hands some years before and had smashed two stocking frames, simple machines that undercut the cost of hand knitting. Workers rose in arms and for two years these Luddites seemed poised to bring the spirit of revolution that had recently transformed the United States and France to the United Kingdom.

In fact, Ned Ludd may never have existed and certainly had much less significance than he was given by the rebellion, but the principles of the Luddites came through clearly and would be seen many times again. There was fear of change. This is a natural human response, because change usually means risk. There was also concern that unskilled workers would lose their jobs—and it happened. This pattern would be repeated in many industries over the years, as technology displaced human beings from jobs that were repetitive, dull, and often dangerous.

While there is good reason not to accept all change for the excitement of the new but rather to assess the associated risks and benefits, the fear of change for its own sake is no sensible argument against alteration. Similarly, although in the short term the loss of low-skill jobs is devastating to the community when it happens, in the long term, provided there is appropriate support for those affected, there is every reason to be positive about this kind of change. Where, for example, mines have shut down and the local communities have been supported to create more modern businesses, few ex-miners would argue for the benefits of being down in the pit. Even fewer would want their children to follow them into such a job.

Similarly, the enhancements that make it easier to learn and remember, for example, should be something we embrace as doing away with unnecessary pain, rather than worrying about the passing of the noble suffering of revision. There's something of the Victorian upper-class concern that labor-saving devices would make servants lazy about ethicists who worry too much about the loss of having to slog through endless repetition to learn facts that will then be churned out in examinations. It would be so much better if those facts could be provided easily to everyone, allowing the examination to concentrate not on rote memory but on interpretation and use. The test would be not a measure of the facts we can remember, but what we can do with those facts, what we can make with the creativity that makes us human.

The President's Council on Bioethics has suggested that in a world of enhanced humans life would be bland and uninteresting. Our poor, improved humans, they suggest, would lead "flat empty lives, devoid of love and longing, filled only with trivial pursuits and shallow attachments." Leaving aside the difficult-to-understand leap that assumes that those with enhanced brains would be any less able to love or to desire, there is definitely something unpleasant about the assumption that our lives would be flat and shallow if we didn't have to do all the grunt work that is involved in the less useful parts of learning.

There is no basis for this assumption, apart from the age-old irritation of those who had to suffer inconvenience that the next generation has managed to avoid. When slide rules were replaced by calculators, everyone said the next generation wouldn't understand math the same way—it didn't happen. When computers came along, many predicted that learning as we know it would collapse. It didn't. Of course, there were new ways of cheating that (for instance) the ability to cut and paste from the Internet would bring, but these were far outweighed by the benefits brought by having a research library at your fingertips, wherever you are in the world.

Fundamentally, those who panic at the thought of enhancing the basic human being miss the point that is the basis for this book. We have been enhancing ourselves ever since our brains developed enough to start making use of the things around us. As Arthur Caplan, a bioethicist at the University of Pennsylvania in Philadephia, has said, all technology is an attempt to transcend human nature. Caplan points out that agriculture, clothes, plumbing, transportation systems—all of these—take us beyond our natural state. Caplan asks the question that is at the heart of the message of *Upgrade Me:* "Do they make us less human? Or are they one possible contender for what it means to be human?"

This is certainly a counter to the metaphysical concern expressed by Bill McKibben in his book *Enough*. McKibben worries that by enhancing our natural abilities we take away the human struggle with our limitations and remove the meaning from our achievements. The point, he argues, of being a musician would be lost on a daughter who had enhanced musical ability, because her piano player mother wanted her offspring to be an even greater pianist than she was herself. The girl would be driven not by the music but her parent's plans. She would never be certain if it was "her skill and devotion or her catalogue proteins that move her fingers so nimbly. . . ."

Yet the same could be said for children brought up in a family of musicians. They, too, have a better genetic start in life than the average child, plus a powerful musical nurturing. I know just such a family of four children, now adults, and they vary from an enthusiastic musician to a young man who has no interest in music at all. The best of the musicians doesn't worry about where her skill comes from. The natural abilities and nurturing didn't make the young man with no interest feel any obligation to take music up. McKibben's imagined pianist's daughter would still require skill and devotion to become a great pianist—to talk dismissively of proteins in this context is bizarre, as any pianist needs the right proteins. And the "programmed" pianist could equally well end up a chef or a social worker.

The picture we get when McKibben worries about us losing our individuality is a one-dimensional view of a person defined by a single characteristic, just because that person has a skill. Of the pianist's daughter McKibben says: "Because . . . one late-twentieth-century woman found solace and meaning in playing the piano, her descendents yea unto the generations are condemned to an ever-deepening spiral of musicality, one that they did not choose and that may haunt them. . . ." What puppets to their abilities these imagined people seem to be.

If we really had the ability to tweak genes to such precision as being able to make someone a better piano player (probably a goal that is much too subtle to be practical, compared with the more attainable possibility of making someone more intelligent or stronger), it would be a gift to the world, not a threat—and in many cases would be ignored and not used, like the vast majority of our capabilities in this short life, unless Kurzweil and friends get their way and our lives are indefinitely extended.

Part of the concern behind the objection to giving a child enhanced abilities is whether parents should be allowed to have this much influence on their children. If we're in favor of choice, how can we allow an enhancement to be "forced" on a child before he or she is even born? Yet parents make a far bigger decision on their children's behalf—whether or not the children should exist at all. By comparison, the decision of whether or not to give them an enhanced memory, more strength, or more intelligence (if these can ever be pinpointed) is relatively trivial.

McKibben also worries about our losing work to the increasingly effective and flexible mechanical workforce. As he puts it: "What, in other words, are we being enhanced for? What does humanity gain if we lose one of our most important defining factors, the work we do?" This is arguably a question that can only come from someone who doesn't hate his or her job, who doesn't spend nine to five waiting for the end of the workday, and who doesn't feel empathy with the name of the restaurant chain T.G.I. Friday's. Automation has been taking over unskilled jobs for many years, and that is going to continue.

As the technology gets more advanced, it will also eat into work that we consider less manual labor and more brain work. At the extreme, we could come to a condition where practically no one has to work. This clearly would require careful handling from the economic viewpoint but doesn't have to be a disaster for humanity. Instead it means that human beings can concentrate on what really interests them. If that happens to be making pottery or writing books, even though computers can produce "better" goods, then there is still the opportunity to do this—and, given the fickle nature of human interest, there is likely to be a market for it.

Life would certainly not have to be one long holiday. It would mean that we could concentrate on what really drives us, makes us human—whether it's the urge to create, to play a sport, to act or play music, or whatever turns us on. For many it will be a wonderful release. How many people, whose lives now consist of working all day to exhaustion, then spending the evening slumped in front of the TV or drinking, could have much more in their lives? It seems extremely patronizing to say that these people couldn't do better without the pressure to bring in the next dollar. Of course some would simply lie back and enjoy—but equally, the freedom from work as we know it could reveal many new talents.

Of course there's a certain nostalgic appeal for the good old days, when we lived by the sweat of our brow. Those were real frontier times, and the attention of robots and nanobots to deal with all the chores will spoil us all, taking away our frontiers by making everything easily accessible? This is surely the weakest of arguments. The frontiers would still be there—they just would have moved. The frontiers that humans face now are the sea and space (and bringing peace to the world)—no amount of effort by robots is going to make those go away as frontiers in the foreseeable future. As for the sweat of the brow stuff, much of that backbreaking, dangerous work is not something anyone will miss. But why should having robots stop us doing the things we want to do?

Surely removing financial pressures will open up our opportunities, not take them away. We won't be left with nothing to do but with a much wider choice. If you enjoy the labor of tilling a field and growing your own food, you can do it, without worrying, as many present-day farmers do, about whether or not you can support your kids. Okay, you could get a perfect carrot out of a nanobot assembler, but this wouldn't take away the sense of achievement of growing one yourself, any more than the fact that I can buy much more beautiful, uniform carrots from the supermarket stops me from growing strangely gnarled and stunted but undeniably interesting carrots in my garden. If you want to build a pyramid in your backyard so you can feel the effort of cutting blocks of stone by hand and hauling them into place, why shouldn't you just because robots are capable of doing it with less effort? This is a nonargument.

It has been pointed out that most visions of a future where technology has transformed human life are dystopias, places of drudgery, pain, boredom, or rebellion. This is true but misses the point of fiction, and particularly science fiction. Science fiction isn't intended to be a prediction of the future; it's a thought experiment, putting human beings in interesting settings and seeing how they react. Utopias are never as interesting as dystopias. It's hard to portray heaven in fiction without it seeming dull, while hell is inevitably decidedly interesting. Dystopias make more interesting fiction—but that doesn't make them a good predictor of the future.

We might be more separated from a romantic idea of nature now than people were in medieval times, but it doesn't stop our lives from being much more pleasant than they used to be. Of course it's dangerous to be blindly optimistic. There will be problems. We will have to be careful that we don't throw the baby out with the bathwater. But the charge that the change implied by our future technological transformation will make us unhappy is a very poor argument.

Another, and perhaps more realistic, concern is the sheer cost of some aspects of further enhancement. Life extension is an obvious problem here. We are already overstretched in providing services for the elderly. The older we get, the more is spent on our health care (which means higher insurance payments for everyone), the more we need support from the family and from the state. If more of us live longer and longer, there is a real danger that a system that is already on the edge of collapse will be pushed into ruin.

Governments around the world are already looking at making the retirement age later in order to get some breathing space. If the life extension technologies described in chapter 2 really do become commonplace over the next twenty years as Aubrey de Grey envisages, won't this make things worse, as more and more people survive into older and older age? Some commentators argue the reverse.

The theory goes something like this. Good life extension doesn't extend the period during which we tend to make heavy claims on health care but rather the period in which we're healthy and capable. When fruit flies and mice have their life extended, they undergo what scientists call compressed morbidity. Rather than having long-drawn-out old ages, the life-extended creatures are fit and well almost to the end of their lives, then quite quickly deteriorate and die. Because of this, not only can governments make retirement age later, reducing the strain on the pension budget, but we can also put off the time when those individuals will make big demands on the medical system.

This sounds great, but there's a feeling of a pyramid (Ponzi) scheme about it. If we get better at preventing early deaths, we won't just put off people getting old and frail; there will also be more people surviving into the future. So when they do eventually become a significant cost, there may be many more of them (or should I say *us*) to cope with. It's arguable that life extension puts off the crisis in the health-care budget but at the same time leaves an even bigger crisis waiting to happen.

Of course, if Ray Kurzweil's dreams of living indefinitely, with deaths largely due to accident (there's no long-term health care required for many accidental deaths), then this problem will never happen. We can just keep putting things off forever. But few believe that Kurzweil's vision is anything more than wishful thinking, and even those who do think he's got a point are pretty sure that it's only the few who will get this "live forever" treatment.

Without doubt, if we did all live indefinitely, it would present us with three broad choices for the long-term future. We would have to either slow down reproduction to a trickle that balanced out accidental death, expand the human race off the Earth, or set an age limit at which we have to gracefully take our leave with euthanasia (this could just be by coming off the life extension program). Even without Kurzweil's ultimate dream, just doubling the average life span is likely to make one of these options necessary. Each might initially seem unacceptable or impractical, but there has to be some limit to the world population, and if we continue to add people at a frantic rate but quit losing them, something will have to be done.

The first is perhaps the closest to a practical solution, but not one that is likely to be politically acceptable. Admittedly China has had a limited child policy in place since 1979, restricting births to one child in the city and two in rural areas, which is still enforced today in the relatively open modern Chinese state. In May 2007 a Chinese businessman was fined 600,000 yuan—a massive $78,000—for breaking the one-child rule. This is not an approach that is likely to find much favor in the West, however, and all the evidence is that it has had limited effect on China's already-falling birthrate.

However, it is possible to imagine some kind of voluntary scheme that would appeal to the same "save the planet" feelings that are played on when we are encouraged to recycle or cut carbon emissions. Perhaps we could reward families for not having many children. Perhaps full life extension treatment would have to be only on offer to those who make such a sacrifice, as the ultimate reward.

The alternative of leaving the Earth is one that science fiction has dangled in front of our noses for many years and one that often appeals to the same wildly optimistic vision of the future that surrounds Kurzweil's "live forever" program, but in reality it is a whole order of magnitude more unlikely. In the years since the Moon landings, the biological science that supports life extension has been totally transformed. Back then, it would have seemed that such medical breakthroughs were even further away than sending colonists into space—but now things are very different. During that same period, space travel has really not advanced at all.

This isn't to say that great things haven't been done in space. Space observatories, GPS, space imaging of the Earth, shuttle missions—there have been plenty of near-Earth developments, and excellent results have been achieved with unmanned probes to Mars and the outer planets, but we have not taken the same strides in manned missions beyond our immediate orbit. We have not been back to the Moon. We have not been to Mars. Although at the time of writing there is talk of taking on these challenges, ever justifying the sort of huge federal expenditure required by the Apollo mission—much more, in fact, to attain the goal of a manned landing on Mars—seems highly unlikely. Many would argue that such massive centralized spending verges on the un-American.

To make the idea of space colonization feasible to reduce

pressure on the Earth's population would require a huge transformation in the way we travel in space. It would require cheap, safe propulsion systems to get the spacecraft out of the Earth's gravity well (perhaps the space elevator described on page 243). It would mean building spacecraft on which hundreds or thousands of people could live for months, protected from radiation and impact, able to exercise so their muscles don't waste away. And it would mean having the technology to terraform a dead planet like Mars, giving it an atmosphere and water to support life.

All of this is not just a matter of getting a little better at what we can do now, of making incremental changes to our capabilities—it is light-years away from our current launches of tiny capsules powered by immense rockets. There is no feasible path from here to there, unless there are some amazing breakthroughs in technology. And that's just to colonize Mars, about the only planet in the solar system that could be made habitable for human beings. Getting to the stars, where there no doubt are many suitable planets, takes us even further into fantasy.

At the moment there are no technologies even on the horizon that would enable faster-than-light travel, an essential to cover the immense distances between the stars. Although some physicists have speculated about mechanisms that could, in principle, allow a ship to break the light speed barrier, they generally involve building vast cylinders of neutron star material or manipulating black holes, "technologies" that are so far beyond feasibility that they might as well say we could lasso a planet from around a distant star and haul it into orbit around the Sun.

When a TV show such as *Star Trek: Enterprise* shows faster-than-light travel being developed in the twenty-second century, it demonstrates phenomenal optimism. It's not that the path from where we are now to having a successful warp drive is a long one, as you might think when looking at the time taken from the first conception of the steam engine back in ancient Greek times to the arrival of real working steam engines in the nineteenth century. In this case there just isn't a path from here to there. There is a huge discontinuity at the point where faster-than-light travel is invented. "Never say never" is the motto of many physicists— but this isn't a concept to bet your life savings on.

Finally comes the least palatable solution, but one that some observers have suggested would become necessary if Kurzweil's full vision were realized. If a large proportion of the population is living indefinitely long lives, then there may come a point when it becomes necessary to intervene to end things. There is a distorted reflection of such a culture in the novel *Logan's Run* by William F. Nolan and George Clayton. In the novel (which is much better than the movie and TV show), the world is run by the young and everyone is expected to submit to euthanasia at the age of twenty-one. The eponymous hero, a "sandman," is tasked with killing those who run, not accepting death; for the rest it is something to be celebrated.

This all seems bizarre and disturbing in the novel, with lives ending when they are just beginning, but if people are living to one thousand or longer it is certainly possible to imagine that many people will want to bring their life to an end at some point. Those who imagine living hugely long lives generally have interesting careers and see a whole life of interest stretching ahead. Things might be less appealing if your vision of the future is one thousand years of cleaning toilets. That's euthanasia on the voluntary level, but equally, if people truly could live forever, is it entirely inconceivable that those who find themselves waiting a thousand years for a promotion or to own a house might begin to feel there has to be a limit?

This is without considering the overcrowding and excess use of resources that are likely to follow from hardly anyone dying. Even without going all the way to imagining enforced euthanasia at a certain age, it is likely that there would have to be legislation to stop people clogging up business. Limited-tenure contracts, for example, for everyone from CEOs to university professors. Living for such a long period is likely to imply repeated changes of career and personal direction, both to keep the individual interested and to make sure the younger people have a chance. It's bad enough being a twenty-year-old when the world is run by the over-forties—imagine dealing with life where most senior executives are entering their second millennium.

With a sub-Kurzweil scenario, where life is extended but not to indefinite levels, population rise might not be such a big problem as may be imagined. Looking back over history, people have tended to have bigger families when life expectancy was poor. In the nineteenth century it wasn't unusual in the West to have ten children, because you knew a fair number would die and you needed all the manual labor you could get. Now populations tend to be falling in the West, except where immigration confuses the picture. There's no need for those big families, and two children or fewer is often the norm.

As might be expected, the champion of "Enough," Bill Mc-Kibben, is highly dubious of the benefits of extended life. His argument, like that against genetic modification, is one of quality of life. We are, he suggests, animals whose defining feature is our awareness of our mortality. It's certainly true that the knowledge that we are going to die has driven a fair amount of our superevolutionary development—ironically enough, those parts that involve preserving life and dealing with illness, the very technologies that now threaten to do away with that driving force.

It would be, McKibben argues, a pointless and wasted existence if we were immortal. This might be true if we literally were unable to die, but even if it were possible to indefinitely extend life, eventually most of us would succumb to accidents, and many ideas of life extension are about doubling our life span rather than getting us to live forever. While no one could sensibly say how they would occupy themselves if they lived for eternity, there are few of us who reach seventy or eighty and, if still healthy, wouldn't want to go on a bit longer. My own father died in his late fifties. Neither he nor the rest of his family would find any objection to a doubling or tripling of his life span.

In the end, living forever would be a voyage into the unknown. It might be after two hundred, or a thousand, or ten thousand years that someone would say, "Enough, I've done it all; I want it to finish," and seek euthanasia. It is almost inconceivable that we could have indefinite life extension without making the option of choosing to end it all legally available. Yet we certainly can't know that any one of us will want to end things at a particular age—and to limit that life span to seventy or eighty because it is "enough" is small-minded indeed.

I have yet to come across anyone who is healthy in mind and body and is having a pleasant life who says, "Okay, I've done all I want to do. I don't want to live anymore." Whether they are twenty or one hundred, the evidence is stacked up against those who proclaim that our natural life span is long enough. The fact that we might eventually lose our drive because we don't have the deadline and threat of death is something we might eventually have to deal with, but I'm prepared to take that risk, and I think most of the rest of the world is, too.

Once more, it's a fuzzy romantic ideal that drives those who crave death at three score years and ten. For some, it's a God-given requirement (even though those biblical ancients did manage to stretch life out toward a thousand). For others, death is a poetic necessity. Without it, says McKibben, "[w]e would be disconnected from the body; even if we still had a container, even if we retained the sound of the beating heart out of some sentimental tic, the body would be much more like a car than a carcass. Just something to carry our brain around in." Setting aside the absurdity of this statement, as the body would still have all its sensory functions that make it much more than just a mode of transport, it's possible to ask, "So what?" Which would you rather inhabit, a car or a dead carcass?

S. Jay Olshansky, whom we met in chapter 2—a gerontologist who has little time for those who believe we can live forever, even though he is positive about some aspects of life extension—was very clear when giving evidence to the President's Council on Bioethics that he did not foresee a problem from population growth caused by anything we achieve in the field of life extension. He asserts that even if Kurzweil's vision of living forever were possible, the growth rate in the population would not be as high as it was during the baby boom that followed World War II.

Total removal of death tomorrow, Olshansky claims, would mean a doubling of population growth between now and the end of the century but nothing more startling. And that's assuming no one would die, which is not going to happen whatever your scenario. Around three people died while you read that previous sentence. There is still a lot of truth remaining in Benjamin

Franklin's remark "[b]ut in this world nothing can be said to be certain, except death and taxes." Any realistic consideration of the impact of life extension is that it will only make a few percentage points' difference in population levels in the medium term.

However you personally regard the possibility of extending life or increasing the intellectual and physical capabilities of human beings, our current unnatural evolution does generate one very real danger that puts us at risk today. Because our technology has become an effective part of us, we can't exist without it. As disaster movies are so fond of showing, it would only take a few days without power and gasoline for our cities to become unsustainable. The very fact that we are so reliant on our technology emphasizes the way we can no longer be sensibly separated from it.

In the end, when assessing the dangers of modifying ourselves, we need to be able answer a simple question—can we do anything to stop it, even if we wanted to? Some, like Bill McKibben, feel that we should be prepared to say, "Stop now. We've got enough. We don't need to be any stronger, or cleverer, or to live any longer." But we don't have a great history as a race of being able to stop something that is within our grasp and appears to deliver personal benefits. Just take a look back in twentieth-century history to prohibition if you think that's an easy step to take.

There have been examples of a rejection of particular technologies in the past, but they have usually been either parochial (as in the Japanese rejection of guns from the seventeenth to the nineteenth century because they weren't honorable—or at least class conscious—enough) or restricted to a particular chemical, rather than a technology as a whole. The most successful example, perhaps, was the banning of CFCs, resulting in a visible shrinkage in the hole in the ozone layer. Certainly we've stopped using one family of chemicals, when many said that would prove impossible, but the technology that made use of those chemicals remains. It's not as if we've stopped using the refrigeration technology that employed CFCs.

Perhaps the best we can hope for in the case of the most dangerous of human changes, like germ line genetic modification, is the level of control we apply to nuclear power. It's not ideal, but it does give us some pause for thought and has a degree of international agreement. There's also a good practical parallel. It is nontrivial to produce nuclear fuel, just as it is not easy to undertake the work necessary to modify human beings at a genetic level. Controls are practical, because it's not an activity anyone is going to do in his or her backyard. At the moment, rules on genetic work vary hugely from country to country. It's arguable we should have a more universal and better-policed attitude to genetic engineering—but this is difficult to maintain when there are strong factions taking opposing views.

It would be contrary to human nature to expect us to be able to shut Pandora's box and pretend we didn't know how to do genetic work or to cut off the benefits. This makes it unlikely that we will be able to avoid selective modifications for (say) more intelligence, either—the incentives are too strong. To ban this when so many are in favor would require a worldwide police state—the cost of the solution would be worse than the problem.

Instead we can hope that the steps are taken carefully, one at a time, with due consideration at each stage. Yes, we will see all of the good things the technologists predict, but probably on a slower time frame. Will we also see the loss of humanity that those who hate this idea suggest will befall us? That is a lot less certain. We can make this work if we are careful. Most people would like children who are stronger and more intelligent and live longer. No one wants children who are less human. It is a balancing act along a fine wire—but it's one that we've had thousands of years of practice at walking, and though our progress is faster than ever we are still keeping up.

It would be foolish to portray our ability to enhance ourselves as wholly positive. When I am not writing books like this, I help large organizations with their creativity. A point I always have to make—sometimes quite forcibly—is that creativity involves risk. You can't do something different, however trivial, without a risk being attached. Sometimes there will be failure, but that's the price we pay for improving things. The only way to avoid risk entirely is to not change anything (and make sure nothing around you changes as well). This doesn't mean that to be creative you have to take blind risks, nor that you should avoid mitigating those risks, but it would be lying to say that they don't exist.

There's a direct parallel between the risks involved in being creative and the risks we take on as a result of changing ourselves and our capabilities beyond our evolutionary norm. In both cases risk is unavoidable, though it can be minimized. In both cases the benefits far outweigh the controlled risk. The idea of some wonderful age in the past, without technology, when everything was more rosy and rural and bucolic is pure romantic fiction. We can't go back, nor would anyone sensible want to. We will go on, though carefully, trying to keep those risks down.

The trouble with the sort of dramatic worries coming out of

biological modification and cyborgs is that such headline-grabbing stories hide the biggest risk from our modifications, which isn't some future possibility but is upon us already. In order to enhance ourselves, we have placed a huge strain on our environment, and all the evidence is that our efforts are now resulting in climate change. This, if anything, is the real monstrous outcome of our upgrading. A very small handful of scientists may still be suggesting that man-made climate change isn't real, but now there is a huge consensus that the problem exists and puts us all at threat.

The subject of this book isn't global warming, so I don't want to spend long on it, but I do want to emphasize just how significant a problem it is. Three news reports in June 2007 all bring out the fact that we have put ourselves in real danger. The first, from the National Academy of Sciences (NAS), revealed that the climate had been changing up to three times faster than the worst predictions. This is particularly worrying because worldwide governmental plans to deal with global warming (when it isn't being entirely ignored) have been based on average predictions of how fast change will occur. The Intergovernmental Panel on Climate Change, which has already been giving out dire warnings on the impact of global warming, has always ignored the worst predictions as too extreme. Yet according to the NAS, the reality is even worse.

The second report was an article in *The New York Times,* written by the secretary general of the United Nations, Ban Ki-moon. Ban points out that the crisis in the Sudanese region of Darfur can be blamed directly on climate change. Over the last twenty years, rainfall in Darfur had dropped by 40 percent. Where once there was water to go around, farmers began to fence off their land, keeping out Arab herders. Tensions rose, and with insufficient food and water for everyone, fighting broke out, leading to slaughter and the displacement of millions.

Finally, in a paper published in *Philosophical Transactions of the Royal Society,* one of the world's top peer-reviewed journals, six leading climate change scientists underlined the dangers posed by global warming. The UK newspaper *The Independent* starkly reported on the paper: "The Earth stands in imminent peril and nothing short of a planetary rescue will save it from the environmental cataclysm of dangerous climate change." The paper's authors suggest that we've come to the point where it simply isn't enough to reduce greenhouse gas emissions, but we need to actively reduce the levels of carbon dioxide in the atmosphere.

If we are to worry about the impact of our upgrading and the technology and manufacturing required to support it, by far the most immediate threat is from climate change, not any of the biological technologies mentioned earlier.

With the very natural wariness of change behind us, we can move on to assess the reality of Human 2.0.

8.
Human 2.0

How beauteous mankind is! O brave new world,
that has such people in't.

—William Shakespeare, *The Tempest*

Thanks to our directed, unnatural evolution we can hugely outstrip the limitations of biology to overcome the opposing forces of nature. There are two primary assertions that my thesis challenges: that we are the same as the original, "modern human" *Homo sapiens* that emerged about one hundred thousand years ago and that we are rushing uncontrollably toward the Singularity, a point in the near future when we will cease to be human.

Each is put in context when we think about our response to those five drivers to upgrade: living longer, becoming more attractive to the opposite sex, being better able to defend ourselves, making the most of our brains, and repairing damaged bodies. The biologists take too narrow a view. We may remain genetically identical to our early ancestors, but this is nothing more than biological "stamp collecting"—driven by classification, rather than the underlying reality. With our unnatural, directed evolution we have picked up capabilities that transform us far more than any changes involved in crossing a species boundary. Version 1.0 of *Homo sapiens* is long gone. As for the Singularity faction, they have focused too much on the future. There is no need to employ guesswork about what will occur, a notoriously inaccurate pastime—instead we can look at what already has happened.

A good example of the narrowly biological view of human nature is the observation made in a serious scientific journal that chimpanzees are more evolved than human beings. To be fair, the author of the article qualifies this statement by putting "evolved" in quotation marks and saying that it's only in one sense that it's true, but the point was still made.

Scientists at the University of Michigan in Ann Arbor have compared 14,000 matching genes in the human and chimpanzee genomes. Of these, 233 of the chimp genes have changed as a result of positive selection—where natural selection appears to have kept the change as giving benefit to the species—as opposed to just 154 of the human genes. "The result overturns the view that, to promote humans to our current position as the dominant animal on the planet, we must have encountered considerable positive selection," the article describes the lead researcher from Michigan as saying.

There are various provisos to the report. It's pointed out that we don't know which genes produced our large brains. "It is possible that the genetic changes underlying brain size are very few," says Jianzhi Zhang, the lead author. And the study could not compare all of the two genomes, as the chimp genome hasn't been sequenced to the same level of detail as the human equivalent. Even so, the article concludes with a comment from Victoria

Horner of the Yerkes National Primate Research Center in Atlanta, Georgia: "We assume that chimpanzees have changed less than us when that's actually not the case."

The problem with this limited pure-biology approach to measuring the degree of change is twofold. First of all, as the report's lead author implies, counts of genes that have been modified are not effective measures of the way an organism has developed. Sheer numbers of genes don't give a useful picture of the complexity of an animal or plant. Like many other relatively simple organisms, the rice plant has significantly more genes than a human being, but even the brightest rice plant is not likely to write any great literature, make a scientific discovery, or, frankly, have any plans for the future. A small number of genes can be responsible for a phenomenally important difference in a creature, with our large brains as a prime example. All genes are not equal.

Second, and most important, to say that in the last 6 million years chimpanzees have changed more than human beings is quite ludicrous, taking a realistic view rather than that of a narrow study of genetics. In that time chimps have, well, carried on doing what chimps still do, with very minor differences. They haven't developed the ability to fly. They can't cross deserts with no water holes for days and live. They can't exist in space (unless we make it possible for them). They can't survive illnesses that should kill them or see what is happening on the other side of the world. Our quasi-evolution through the capabilities of our brains leaves the chimpanzee on the evolutionary starting block.

So we come to the Singularity. (Does it deserve a capital letter? Probably not, but I'll preserve the convention that is usually employed.) The very name gives away the grandiose way the enthusiasts view their idea. In math and physics, a singularity is something that involves the infinite; it's a phenomenon that breaks the bounds of the universe as we understand it. In a physical singularity, the usual laws of nature don't exist. That's pretty awesome. By comparison, the Singularity as a vision of a future where we use technology to become something more than human lacks that sense of the ultimate, though admittedly it does imply a discontinuity in our nature. (In fact, Discontinuity would have been a much more apt title for the idea, but it doesn't sound as impressive as Singularity.)

In many ways, the concept of Singularity, as a vision of us (that is, conventional Human 1.0s) being replaced, whether we like it or not, by a more-than-human biological-technological hybrid is self-defeating. It is true that we will eventually adopt brain enhancements (see page 227). We will use nanotechnology to support our biological systems. And it's even true that the capabilities of the implants and support technology we use will far outstrip our own biological functionality. But this doesn't mean we have to lose control. We will make sure that we keep the upper hand.

Take two commonplace technologies from the twentieth century, the car and the computer. Physically the car is vastly more powerful than a human being. Although horsepower is no longer directly linked to the strength of individual horses, it gives a guide to the sort of effort a car generates. It's not unusual for an automobile engine to achieve 200 horsepower—some deliver well over 500. Compared to the potential power of the human body, this is immense. They may tow trucks in *The World's Strongest Man,* but you'll never see one of these muscle-bound contenders taking on a truck pull for pull. It would be a walkover.

Yet despite this power that far outstrips our biological capabilities, we control our cars. Some might argue this isn't true, that automobiles kill us and ruin the countryside, but even then it is human beings who make this happen. For all their power, cars and trucks answer to our command.

Similarly, the cheapest personal computer can far outstrip the human brain for sheer speed of calculation. Don't get me wrong—brains are awesome. Brains are hugely complex mechanisms and in this complexity outstrip any existing computer. But the processes of the brain are slow. They combine chemical reactions with electrical signals, and impressive though this mechanism is, chemical processes are sluggish when compared with electronics. A personal computer can undertake billions of calculations in the time even the best human mathematician can only work through a handful. Yet we tell the computers what to do. They haven't taken over.

We might sometimes think our PCs have minds of their own, when they refuse to do what we want—but that's just the (human) programmer getting something wrong. For all their power, computers are tools. The same will be the case with the additions we make to our human capability. They may be able to work faster than brain cells, but they will still be under our control, and we will make sure this is the case.

My main problem with the Singularity prophets, though, is not a vision of "normal" human beings being sidelined by a cyborg—it is the shortsightedness of the suggestion that "sometime soon" we will be so transformed that we are no longer human. This transformation, as we have seen, has been under way for tens of thousands of years, giving us capabilities far beyond that of the biological human.

"Our version 1.0 biological bodies are likewise frail and subject to a myriad of failure modes, not to mention the cumbersome maintenance rituals they require. . . . The Singularity will allow us to transcend these limitations of our biological bodies and brains," comments Ray Kurzweil, chief enthusiast for the Singularity. But don't we already transcend the limitations of our biological bodies and brains every time we step into a car or a plane, use antibiotics, or surf the Internet? This isn't a new paradigm; it's a summary of human history.

To be fair, those who prophesy that the Singularity is imminent are using some impressive data. The basis for their predictions is the exponential growth in our technologies. Exponential growth is just a change where the bigger the thing growing is, the faster it grows. The result is a graph of growth that starts slowly and then shoots up toward the end. Perhaps the best-known example of exponential growth in technology is Moore's law, the rule of thumb that describes the way the size of commercial computer processors grows with time.

In 1965 Gordon Moore, then the head of R & D at Fairchild Semiconductor, and later to found Intel, predicted that the number of transistors in a processor, a crude measure of its power, would double every year. His observation at the time was based on only a few years' data, and he later revised this to doubling every two years—the reality has hovered between his two predictions for over forty years now.

It might sound as if something doubling every year, or every two years, hadn't got the element of explosive change that is implied by exponential growth, because there appears to be something stable and regular about the "every two years" part. But each time the starting point is twice as big as the last. The bigger the transistor count, the faster it is growing. Starting with only four transistors, processor size would grow by four in two years. Starting with a billion, it grows a billion.

Exponential growth doesn't just explode off the graph; it rapidly escapes our comprehension. We find it difficult to anticipate the impact of exponential growth. According to legend, the peasant who invented the game of chess was offered a reward by a grateful emperor. (Another version of the legend has an Indian king playing chess against a sage, who turns out to be Krishna in disguise.) The peasant asked for what seemed a trivial reward. One grain of rice for the first square of the board, two for the second, four for the third, and so on, doubling up the amount for each of the sixty-four squares. The emperor agreed, only to discover to his horror that before long he had used up every bit of rice in his empire. To complete the sequence up to the sixty-fourth square would require around 37,000,000,000,000,000,000 grains of rice.

Exponential growth is difficult to get your head around, because the natural assumption is one of linear change. This is simple straight line change. We assume that the way things alter in the future will be much as it was in the past. This seems to be the inbuilt human understanding of the world—we expect things to be linear. Our natural tendency is to forecast what is going to happen based on a linear change that approximates to the rate (and direction) things have changed before.

This produces a tendency to poorly predict the future because we expect "more of the same." A good example from the software world was the launch of Windows 95. Back in 1995, all the successful electronic networks were privately owned and provided limited, self-contained services like America Online and CompuServe. So Microsoft, in launching Windows 95, produced its own private network, MSN. Even though the Internet was already well established, a savvy organization such as Microsoft wasn't able to spot that this would be the electronic communication vehicle of choice and that the private networks would rapidly wither to become adjuncts to the Net.

Because of this assumption of the linear, anything that is undergoing exponential growth is likely to far exceed our expectation. This is the primary argument for the importance of Singularity. As the key technologies that contribute to the concept—biotechnology, nanotechnology, and computing—all shoot off up the exponential curve, they will reach a point where we won't be able to keep up, and a new version of human, empowered by these technologies, will emerge.

Kurzweil points out how scientists tend to be pessimistic, or at least careful with their predictions, and accordingly are often left behind by reality. However, there is a note of caution to be raised when jumping onto the exponential bandwagon. Although we know that technology is moving ahead at this regularly doubling rate and this means linear predictions are wrong, it doesn't mean that everything will happen with explosive suddenness. Remember that classic example of getting it wrong that we've already mentioned, the movie *2001: A Space Odyssey*.

When Stanley Kubrick and Arthur C. Clarke came up with *2001* back in 1968, they were looking thirty-three years into the future. By the year 2001, they expected that Pan Am would be operating shuttles to a huge space station. That there would be regular manned trips to the Moon. That we would be communicating on huge, full-screen, full-motion video phones (provided by Bell). And that a manned probe, equipped with a self-aware computer that could converse in perfect English, could be sent to the moons of Jupiter. All this they set in our past.

Over those thirty-three years, information technology may have moved on in leaps and bounds (though we still can't make HAL), but the technology to put human beings into space didn't change all that much. In the years before 1968 it had grown intensely, but then it hit a stop. Some technological changes need a specific breakthrough to trigger a period of growth, and the next trigger just hasn't happened yet with space travel. The same may be true of nanotechnology (for example). While everything seems to be progressing in good exponential fashion at the moment, we could soon hit a barrier with no obvious way forward.

Kurzweil's exponential predictions have the same problem as "natural feeling" linear predictions. They are still based on extrapolating the past into the future. Some aspects will be true. It seems likely that Moore's law will hold good for a good number of years to come, though even this must come to an end. Increasing speed involves increasing miniaturization, and physics suggests that there are boundaries beyond which this becomes first impractical, then impossible. At some point computing architecture will have to deal with single electrons at a time, and beyond that there is nowhere to go.

This doesn't mean processor power won't keep increasing for many years (at the time of writing, Intel is predicting fifteen-plus more years before having to make a significant change of tack). Manufacturers have shown themselves to be supremely clever when it comes to finding different ways to cram more into the same space or to change the topology of the chip so that more space becomes available, for instance by expanding the structure vertically as well as horizontally—but limits will be hit eventually.

Worse still, though, the software that can run on a computer is inherently limited. As long ago as the 1930s, Alan Turing, the mathematical genius at the heart of the wartime Bletchley Park center that broke the secrets of the German Enigma coding machines, proved that there were some problems that can never be resolved by a computer. And there are many more that, though theoretically soluble, would take the best computer longer than the entire lifetime of the universe to reach an answer to. There may be an answer in the much-vaunted concept of quantum computers, which use individual quantum particles as bits within the device and can, in principle, crack some problems that would take a conventional computer longer than the lifetime of the universe to solve—but these are a totally different technology, jumping onto a new curve of prediction.

It's not, then, that the circumstances some think will lead up to Singularity are impossible, but the predictions of how it is going to happen (like pretty well every science and technology prediction) are likely to be wrong. Ray Kurzweil puts an estimate on the arrival of the Singularity in the form of machine intelligence supplanting our own at a date in the 2040s. His argument is that by then (given expectations of exponential growth) the price of a PC would buy you the equivalent computing power of all the brains in the world—so the electronic intelligence created will be vastly more powerful than all current human intelligence.

There are, of course, some shaky assumptions in that statement. It is assumed that having the equivalent computing power means having the same level of intelligence, a very doubtful step to take. The idea behind the comparison is that the computing power needed to simulate the brain's activity is an appropriate equivalent to the brain itself, but this is questionable. You could imagine a future version of a 3-D IMAX movie that is so good a simulation that there is no visual difference between being somewhere and experiencing the movie—but that doesn't make them the same thing. You still can't walk through the screen and interact with the objects you see, however good the simulation.

The technological prophets of the future also assume that we will continue on the same technical trajectory, rather than abandoning what we are doing at the moment—probably true, but not certain. And it is accepted without question that we will allow technology to progress unchecked, something that hasn't happened with other technologies in the past. Right now, stem cell research, for example, an essential building block toward the Singularity vision, is regularly blocked in many countries, including the United States.

A good illustration of the divide between the excitement of future vision and the more everyday practical reality that could throw a spanner in the works of the Singularity is the way that robotics hasn't really come on the way many would have expected. Ray Kurzweil lists in his book *The Singularity Is Near* page after page of robotic successes, but it is a rose-tinted picture of the robotic reality, one that is easily accepted because our fictional world is populated so strongly with robots that are near to human—a deceit that causes strange misperceptions.

A good example of this is the use of robots in the TV show *Buffy the Vampire Slayer*. Provided you accept the main fantasy premise of the show—that magic and vampires exist—the story lines largely build on this premise in a logical fashion. But on occasions there are humanoid robots in the show (built by a student) that are only distinguishable from real humans by occasional strange behavior. This really grates, because these robots are so far beyond our current-day technology. We can't even build a robot that moves like a human being or a software package that can converse for long without hitting a difficulty. So, strangely, in a world populated with vampires and demons, the robots are the most far-fetched part of the storyline.

A more accurate picture of the robot world than is given by *The Singularity Is Near* can be gained by looking at the sort of work under way in centers of robotic excellence like Carnegie Mellon University and MIT. Robots are being produced that, in principle, can do all kinds of good things. In practice, though, not only are they frustratingly limited, but also they keep going wrong. In 2004 DARPA ran a challenge race. The entrants in the race, with a prize of $2 million at stake, had to produce a robotic vehicle that could traverse a 132-mile desert course in Nevada.

Compared to many of the challenges facing roboticists, this seemed a trivial task. The robotic cars didn't need to be hugely intelligent or to do anything particularly creative; they just had to get from A to B (admittedly, passing through a number of way stations along the way). The manager from DARPA, Jose Negron, booked a sixty-five-hundred-seat ballroom at Buffalo Bill's Resort & Casino as the venue for the awards ceremonies for the winners. But it was to prove a lonely event. None of the competitors came close to the finishing line.

The best of the bunch was Sandstorm from Carnegie Mellon, which managed 7.4 miles before catching on fire. One of the other competitors fell over on the start line. Admittedly, the following year five robots completed the course, but the winner only managed an average of 19 miles per hour—and it still could easily have failed. It is very tempting to underestimate the complexity of getting a sophisticated system like a robot to work effectively. Just imagine, for instance, we've developed the nano-scale intelligent robots that Kurzweil envisages injecting into our bodies to fix problems and beef up the brain. Can you imagine one getting halfway down a vein, then bursting into flames? Or suddenly stopping because it has the nano equivalent of the Windows blue screen of death?

The software for complex robots seems pretty well impossible to debug entirely, and the mechanical parts are always hitting against practical problems that make apparently simple tasks unbelievably difficult. One way of looking at the challenge of effective robotics is that despite making PC software for over twenty years, Microsoft still can't give me software where I can type a document, keep a diary, surf the Internet, and listen to some music without something going wrong on a daily basis. And we expect robots with artificial intelligence that's better than our own to be around in a few years' time? I'm not holding my breath.

Take another example of a sudden stop in a growth curve— the speed of human travel. Bill McKibben points to science writer Damien Broderick's description of the way our speed of travel has increased over time (a development that was driven, incidentally, by our superevolutionary transformation of ourselves, but that's the story of this book). For millions of years we were restricted to walking. Over thousands of years we got a little faster by using donkeys and horses. Just two hundred years ago the steam train arrived, followed by automobiles, prop planes, and jets. According to Broderick, "By 1953, not even the Air Force technologists could believe what the trend curves were telling them: that within four years they would have achieved speeds great enough to lift payloads into orbit."

Despite that disbelief, the curve went on, with *Sputnik* going into orbit on schedule in 1957, and twelve years later man was on the Moon. The idea of this example is to show how exponential growth can transform things in a shocking way—and it's a great example up to a point. Yet in its use of carefully selected data it is also extremely misleading.

To start with, the world didn't end with the introduction of space flight. Bear in mind that almost as much time has now passed since *Sputnik* was launched in 1957 as had gone by between the Wright brothers' first flight and that satellite launch that triggered the space race. Even with linear growth we would by now expect to be crossing the solar system with ease—with exponential growth, we should be hopping to the stars after fifty years had passed. It hasn't happened. Our space vessels are no quicker than they were in the Apollo days.

Look a little deeper, and things are even worse than that as far as the growth curve goes. Broderick's comparison of speeds of travel puts apples alongside oranges. It begins as a description of mass travel speed—the rates that ordinary people can achieve—but ends with specialist travel speed. Very few people have become astronauts. If you stick to the curve of mass travel, something even more remarkable has happened. From the Wright brothers' handful of miles an hour we did see immense jumps with the jet plane and then with the supersonic Concorde at around 1,300 miles an hour. But since Concorde went out of service, our fastest mass travel speed has dropped to less than half of its previous value. Not only are we failing to ascend an exponential curve, or even a linear curve, but we also have actually plummeted backward, with no real contender in sight to reverse this.

It would be foolish to ignore the impact of exponential growth, but it is equally foolish to be so selective of data that you can look at an example like the increasing speed of human travel and not realize that this demonstrates that it is equally possible to hit sudden plateaus and even reverses. We can't assume that things will carry on as they have been up to now.

Is the Singularity coming? Maybe. Certainly there will be immense unexpected changes ahead. But to expect there to be a point in the future when we make the move away from the original, version 1.0, human being is intensely shortsighted. Though genetically relatively unchanged from the earliest *Homo sapiens,* we are in reality as different from them as they were from their predecessors. Far more so, in fact. The Singularity enthusiasts and others argue we will lose our humanity. Bioethicist George Annas of Boston University has suggested that there should be a global ban on genetic modification for this reason. Yet I would argue that it is the pursuit of change, our ability to constantly enhance ourselves, that *makes* us human.

Rather than bemoan our technological capabilities, we should be celebrating. Human 2.0 is already here. Take a look in the mirror.

Notes

Introduction

2   The observation that our fragile skin must have an "obvious evolutionary advantage" is from Julian Robinson, *The Quest for Human Beauty* (New York: W. W. Norton, 1998).

3   Richard Dawkins uses the metaphor of evolution as a "blind watchmaker" in Richard Dawkins, *The Blind Watchmaker* (London: Penguin, 1988).

1. Beyond Biology

8   Bromhall's theory of the origins of the human form and consciousness is described in Clive Bromhall, *The Eternal Child* (London: Ebury Press, 2003).

2. Cheating Death

17   The quote from the Monk of Saint Gall's life of Charlemagne is taken from the *Medieval Sourcebook* at www.fordham.edu/halsall/basis/stgall-charlemagne.html.

19   Details of medieval armor are from Charles Ffoukles, *Armour and Weapons* (Yardley, Pennsylvania: Westholme Publishing, 2005).

24   The evidence of domesticated fire in South Africa is referenced on the UNESCO World Heritage Centre Web site at http://whc.unesco.org.

24 Doubts about the African fire sites and the existence of the site in Israel are referenced in *NewScientist,* April 29, 2004.

30 Information on the history of alchemy is from Peter Marshall, *The Philosopher's Stone* (London: Macmillan, 2001).

31 Isaac Newton's work in alchemy is described in detail in Michael White, *Isaac Newton: The Last Sorcerer* (London: Fourth Estate, 1997).

32 Roger Bacon's ideas on alchemy and secrets are taken from Roger Bacon, *Letter Concerning the Marvelous Power of Art and of Nature and Concerning the Nullity of Magic* (Kila, Montana: Kessinger Publishing, 1997).

34 For more on Roger Bacon and his writing see Brian Clegg, *The First Scientist: A Life of Roger Bacon* (London: Constable & Robinson, 2003).

35 The observation that in 1994 no eight-year-old girls died in Sweden is from Armand Marie Leroi, *Mutants* (London: HarperCollins, 2003).

35 Aubrey de Grey's suggestion that the first person to reach one thousand may already be sixty is made in a number of places, including an article he wrote for the BBC, "We Will Be Able to Live to 1,000," at http://news.bbc.co.uk/1/hi/uk/4003063.stm.

37 Ray Kurzweil's ideas on extending human life indefinitely are described in Ray Kurzweil and Terry Grossman, *Fantastic Voyage* (London: Rodale International, 2005).

37 The study showing the effect of HGH on some aspects of aging was D. Rudman et al, "Effects of Human Growth Hormone in Men over 60 Years Old," *The New England Journal of Medicine* 323, no. 1 (1990): 1–6.

38 Kurzweil puts the decade of the 2030s as the time when "the nonbiological portion of our intelligence will predominate" in his article "Human Life: The Next Generation" in *NewScientist,* September 24, 2005.

41 The paper on combating aging from Aubrey de Grey, Bruce Ames, Andrzej Bartke, and Judith Campisi is "Time to Talk SENS: Critiquing the Immutability of Human Aging," *Annals of the New York Academy of Sciences* 959 (2002): 452–62.

42 Aubrey de Grey's seven deadly things are detailed on his Web site, at www.sens.org.

50 Ian Wilmut's suggestion to use cloning technology to make healthy embryos for IVF is described in Ian Wilmut and Roger Highfield, *After Dolly: The Uses and Misuses of Human Cloning* (London: Little Brown, 2006).

52   Details of Ashanti DeSilva's case are taken from Ramez Naam, *More Than Human* (New York: Broadway Books, 2005).

55   Michael Rose's assertion that aging was not a hot area of study in 1976 is taken from Michael R. Rose, *The Long Tomorrow* (Oxford: Oxford University Press, 2005).

56   Examples of earlier twentieth-century theories on the causes of aging are also taken from *The Long Tomorrow.*

57   The assertion in 2002 that there are no therapies that will put off aging is described in S. Jay Olshansky, L. Hayflick, and B. A. Carnes, "No Truth to the Fountain of Youth," *Scientific American,* June 2002.

57   S. Jay Olshansky's comments on the unlikely nature of predictions of immortality were made in an article he wrote for the BBC, "Don't Fall for the Cult of Immortality," at http://news.bbc.co.uk/1/hi/uk/4059549.stm.

58   Richard Miller's observations of mutations in mice causing longer life is from "The Incredibles" in *NewScientist,* May 13, 2006.

59   The contribution of free radicals to aging is covered in H. K. Biesalski, "Free Radical Theory of Aging," *Current Opinion in Clinical Nutrition and Metabolic Care* 5, no. 1 (2002): 5–10.

65   Ian Wilmut's assertion that Dolly the sheep's postmortem revealed nothing unusual is taken from *After Dolly.*

73   The Institute for the Future's views on genetic modification to deal with different environments are taken from an article by John Thackara, "Putting the Future in Perspective," *RSA Journal,* June 2006.

3. Cosmetic Charisma

75   The estimate for 2010 market size for noninvasive face-lifts and facial rejuvenation is taken from a 3i press release, at www.3i.com/media/press-releases/ulthera_131205.html.

76   The estimate for the global cosmetic market is taken from L'Oréal's *2005 Sustainable Development Report.*

77   The deduction of the dates of starting to wear clothes based on louse DNA and dates of twenty-seven thousand years ago for weaving and forty thousand years ago for needles are from John Whitfield, "Lice Genes Date First Human Clothes," *Nature Online News,* at www.bioedonline.org/news/news.cfm?art=436.

82   Descriptions of early garments, cloth, and weaving are from Elizabeth Wayland Barber, *Women's Work: The First 20,000 Years* (New York: W. W. Norton, 1995).

83   Hilaire Hiler's speculation about the uses of clothing is from Hilaire Hiler, *An Introduction to the Study of Costume* (London: W & G Foyle, 1929).

90   Information on the skin and tanning is taken from Michael F. Holick, *The UV Advantage* (New York: iBooks, 2003).

91   The example of an Iron Age tattoo is described in *Women's Work*.

91   Information on transformations of the body for attraction is taken from Julian Robinson, *The Quest for Human Beauty* (New York: W. W. Norton, 1998).

92   The suggestion that body painting was no longer practiced in the British Isles by the Norman Conquest comes from Richard Corson, *Fashions in Makeup from Ancient to Modern Times* (London: Peter Owen, 2003).

96   The formulation of ancient Egyptian lipstick and other lipstick information is from Rita Johnson, "What's That Stuff? Lipstick," *Chemical & Engineering News* 77, no. 28 (July 12, 1999): 31.

99   For more on the history of makeup, see Corson, *Fashions in Makeup from Ancient to Modern Times.*

100   Information on the history of cosmetic surgery is from Elizabeth Haiken, *Venus Envy: A History of Cosmetic Surgery* (Washington, D.C: Johns Hopkins University Press, 1999).

101   The importance of symmetry in attractiveness is described in "Sex & the Symmetrical Body," *NewScientist,* April 22, 1995.

104   The experiment using photographs with increased pupil size is demonstrated in Desmond Morris, *Manwatching* (London: Jonathan Cape, 1977).

106   Roger Bacon's advice on diet is from Roger Bacon, *Letter Concerning the Marvelous Power of Art and of Nature and Concerning the Nullity of Magic* (Kila, Montana: Kessinger Publishing, 1997).

110   The quote about body modification making a statement is taken from Mike Featherstone (ed.), *Body Modification* (London: Sage Publications, 2000).

110   Paul Sweetman's observation on the reasons for light body modification is taken from "Anchoring the (Postmodern) Self?" in *Body Modification.*

4. The Strength of Ten

127   Information on Roger Bacon and gunpowder is from Brian Clegg, *The First Scientist: A Life of Roger Bacon* (London: Constable & Robinson, 2003).

129   The suggestion that genetic modification of athletes is one step too far is from Bill McKibben, *Enough: Staying Human in an Engineered Age* (New York: Henry Holt, 2003).

131   The different awards for challenges on Mars are described in Charles S. Cockell, *Space on Earth* (London: Macmillan, 2006).

137   The varying attitudes to the dog in different cultures and times are described in Keith Thomas, *Man and the Natural World* (London: Penguin, 1984).

139   Information on guide dogs was provided by the Guide Dogs for the Blind Association Web site, at www.gdba.org.uk.

143   Details of Edison's battle with Swan over inventing the incandescent lightbulb and on the development of artificial light are from Brian Clegg, *Light Years: An Exploration of Mankind's Enduring Fascination with Light* (London: Macmillan, 2007).

145   The Persian flying adventure of Kai Kawus and much of the information on the history of flight is from Riccardo Niccoli, *History of Flight* (Vercelli, Italy: White Star, 2006).

146   Roger Bacon's description of the use of lenses by Julius Caesar is from Roger Bacon, *Letter Concerning the Marvelous Power of Art and of Nature and Concerning the Nullity of Magic* (Kila, Montana: Kessinger Publishing, 1997).

146   The Taoist tradition of becoming a *hsien* who can fly in the heavens is described in Peter Marshall, *The Philosopher's Stone* (London: Macmillan, 2001).

146   Roger Bacon's assertion that it is possible to fly using a machine is in *Letter Concerning the Marvelous Power of Art and of Nature.*

148   Details of the Bognor Regis International Birdman competition are from their Web site, at www.birdman.org.uk.

153   Information on the PoweriZer is from www.powerizer.com.

153   Information on the SpringWalker is from www.springwalker.com.

154   Information on DARPA's Exoskeletons for Human Performance Augmentation program is from the DARPA Web site, at www.darpa .mil/dso/thrust/matdev/ehpa.htm.

157   Details of the BLEEX exoskeleton are found in H. Kazerooni, "The Berkeley Lower Extremity Exoskeleton Project," *Proceedings of the 9th International Symposium for Experimental Robotics,* Singapore, June 2004.

157   Powered suits for the elderly available from 2008 were reported in *The Japan Times,* April 18, 2007.

158   The artificial muscles are described in Ebron et al, "Fuel-Powered Artificial Muscles," *Science* 311, no. 5767 (2006): 1580–83.

5. The Deadliest Weapon

164   Pinker's ideas on the origins of language are described in Steven Pinker, *The Language Instinct* (New York: William Morrow, 1994).

168   The significance of the written word is described in Nicholas A. Basbanes, *Every Book Its Reader* (London: Harper Perennial, 2007).

175   Numbers of periodicals published in California in 1859 are taken from Horace Greeley, *An Overland Journey from New York to San Francisco in the Summer of 1859* (Lincoln: University of Nebraska Press, 1999).

175   Eadweard Muybridge's bookselling in San Francisco is described in Brian Clegg, *The Man Who Stopped Time* (Washington, D.C.: Joseph Henry Press, 2007).

187   Einstein's quote about sitting on a stove and relativity comes in many variants (some "put your hand on a stove for two minutes," et cetera)—this is the original, from his abstract "On the Effects of External Sensory Input on Time Dilation," *Journal of Exothermic Science and Technology* 1 (1938).

189   The suggestion that 10 percent of U.S. university students use Ritalin or other prescription stimulants comes from "The Incredibles," *NewScientist,* May 13, 2006.

191   Improved memory response by students at MIT after taking choline is described in Beverly Potter and Sebastian Orfali, *Brain Boosters: Foods & Drugs That Make You Smarter* (Berkeley: Ronin Publishing, 1993).

192   Eric Kandel's life and work on understanding memory are described in Eric Kandel, *In Search of Memory* (New York: W. W. Norton, 2007).

194   The experiments by Fortunato Battaglia using transcranial magnetic stimulation on mice are described in "Magnets Bolster Neural Connections," *NewScientist,* May 26, 2007.

194   The suggestion that new-grown neurons could be involved in memory formation comes from Shaoyu Ge et al, "A Critical Period for Enhanced Synaptic Plasticity in Newly Generated Neurons of the Adult Brain," *Neuron* 54 (2007): 559–66.

195   More information on Theodore Berger's work on brain implant chips can be found at www.neural-prosthesis.com.

202   Information on Roger Bacon is from Brian Clegg, *The First Scientist: A Life of Roger Bacon* (London: Constable & Robinson, 2003).

202   Roger Bacon's speculation on seeing distant objects through lenses is taken from Roger Bacon, *Letter Concerning the Marvelous Power of Art and of Nature and Concerning the Nullity of Magic* (Kila, Montana: Kessinger Publishing, 1997).

202   Roger Bacon's description of the optical principle of the telescope is taken from Roger Bacon, Robert Belle Burke (trans.), *Opus Majus* (Kila, Montana: Kessinger Publishing, 1998).

202   More on the Digges telescope and the development of telescopes in general can be found in Patrick Moore, *Eyes on the Universe: The Story of the Telescope* (London: Springer-Verlag, 1997).

205   The experiment relaying images from a cat's retina onto a screen is described in Garrett B. Stanley, F. F. Li, and Y. Dan, "Reconstruction of Natural Scenes from Ensemble Responses in the Lateral Geniculate Nucleus," *Journal of Neuroscience* 19, no. 1 (1999): 8036–42.

205   The investigation of direct brain-to-brain communication is described in H. Hoag, "Neuroengineering: Remote Control," *Nature* 423 (2003): 796–98.

207   More details of Kevin Warwick's "cyborg" experiments are available from his Web site, at www.kevinwarwick.com.

6. Body Shop

216   A good example of the limits of early medicine is the work of the medieval doctor Paracelsus, covered in Philip Ball, *The Devil's Doctor* (London: William Heinemann, 2006).

217   Nicholas Culpeper's move of herbal medicine from folklore to accepted medical practice is described in Benjamin Woolley, *The Herbalist* (*Heal Thyself* in the United States) (London: Harper Perennial, 2005).

218   The origin of the germ theory is described in John Waller, *The Discovery of the Germ* (Cambridge: Icon Books, 2004).

218   Information on John Snow's discovery of the mechanism for spreading cholera is from Sandra Hempel, *The Medical Detective* (*The Strange Case of the Broad Street Pump* in the United States) (Cambridge: Granta, 2006).

218   For more on the discovery of penicillin see Eric Lax, *The Mold in Dr. Florey's Coat* (New York: Henry Holt, 2004).

219 The discovery and gradual understanding of the mechanism of aspirin is described in Diarmuid Jeffreys, *Aspirin: The Remarkable Story of a Wonder Drug* (London: Bloomsbury, 2005).

223 Information on Alzheimer's disease is taken from the Alzheimer's Society Web site, at www.alzheimers.org.uk.

224 The UCSD experiment using nerve growth factor to slow the onset of Alzheimer's in described in Tuszynski et al, "Nerve Growth Factor Gene Therapy for Alzheimer's Disease," *Journal of Molecular Neuroscience* 19, nos. 1 and 2 (2002): 207.

225 Basic information on the brain is from Brian Clegg, *Studying Creatively* (London: Routledge, 2007).

225 The extension of deep brain stimulation to cluster headaches is described in D. Black et al, "Small Study Suggests Promising Role for Deep Brain Stimulation Surgery to Treat Intractable Cluster Headache," at www.mayoclinic.org/news2007-rst/4044.html.

226 Further details of the methodology of deep brain stimulation are from the National Institute of Neural Disorders and Stroke Information page, "Deep Brain Stimulation for Parkinson's Disease," at www.ninds.nih.gov/disorders/deep_brain_stimulation/deep_brain_stimulation.htm.

226 Details of use of deep brain stimulation for tremor and Parkinson's disease are taken from National Institute for Health and Clinical Excellence publications IPG019 (2003) and IPG188 (2006).

227 Robert Heath's pioneer work on the impact of stimulation on the human brain is described in R. G. Heath, "Electrical Self-stimulation of the Brain in Man," *American Journal of Psychiatry* 120 (1963): 571–77.

229 The mapping process for phosphenes using Dobelle's technology is described in Stephen Kotler, "Vision Quest," *Wired,* September 2002.

232 The paper describing the experiment giving direct control from the brain to electronic devices is L. R. Hochberg et al, "Neuronal Ensemble Control of Prosthetic Devices by a Human with Tetraplegia," *Nature* 442 (2006): 164–71.

233 The experiment using monkeys' premotor cortexes to predict intended motion is described in G. Santhanam et al., "A High Performance Brain-Computer Interface," *Nature* 442 (2006): 195–98.

7. Monsters and Mutants

241 Michael Crichton, *Prey* (New York: HarperCollins, 2002).

242 For more on nanotechnology, nanobots, and assemblers, see K. Eric Drexler, *Engines of Creation* (New York: Bantam, 1986).

243   Bradley Edwards's description of a space elevator was in an interview for Space.com at www.space.com/businesstechnology/technology/space_elevator_020327-1.html.

249   The BT trio's description of posthuman evolution is from Ian Pearson, Chris Winter, and Peter Cochrane, *The Future Evolution of Man* (Ipswich, U.K.: BT Labs, 1995).

251   George Annas's suggestion that modified humans might want to enslave or slaughter the nonmodified comes from his speech at the Beyond Cloning conference, Boston University, 2001.

252   The figure of two-thirds of children under five dying in 1750 is given in P. Razzell and C. Spence, "Social Capital and the History of Mortality in Britain," *International Journal of Epidemiology* 34, no. 2 (2005): 477–78.

252   The assertion of the length of life of a seventy-year-old in 1900 is taken from L. Hayflick, "How and Why We Age," *Experimental Gerontology* 33, nos. 7 and 8 (1998): 639–53.

252   The difference in increase in life expectancy between the developed and developing world is from R. Guest, "Getting Better All the Time," *The Economist,* November 8, 2001.

257   The quote from the President's Council on Bioethics on leading flat lives is taken from "The Incredibles," *NewScientist,* May 13, 2006.

258   Arthur Caplan's assertion that our technology may be a contender for what it means to be human is taken from "The Incredibles."

258   Bill McKibben's objection that genetic enhancements stop us from being human individuals comes from Bill McKibben, *Enough: Staying Human in an Engineered Age* (New York: Henry Holt, 2003).

262   The observation that most science fiction visions of the future are dystopias is in McKibben, *Enough.*

263   The argument that life extension will benefit countries that are struggling to pay for their elderly is made in Ramez Naam, *More Than Human* (New York: Broadway Books, 2005).

266   Information on *Star Trek: Enterprise* from the official Web site, at www.startrek.com/startrek/view/series/ENT.

269   Bill McKibben's doubts about the benefits of extended life are from McKibben, *Enough.*

270   S. Jay Olshansky's assertion that longer life won't cause population growth problems is taken from "Duration of Life: Is There a Biological Warranty Period?"—his testimony before the President's Council on Bioethics (2002).

270   The assertion of the number of people dying is based on the United Nations' World Population Prospects, 2006 database, at http://esa .un.org/unpp.

271   Benjamin Franklin's comment on death and taxes is taken from a letter to Jean Baptiste Le Roy, dated November 13, 1789.

274   The National Academy of Sciences study showing global warming to be faster than anticipated is Michael R. Raupach et al, "Global and Regional Drivers of Accelerating CO_2 Emissions," *Proceedings of the National Academy of Sciences of the United States of America* 104, no. 24 (June 2007): 10288–93.

275   The article on the Royal Society paper on climate change is from *The Independent* (London), June 21, 2007.

8. Human 2.0

278   The article describing chimps as more evolved than humans is Michael Hopkin, "Chimps Lead Evolutionary Race," *Nature* 446 (2007): 841.

282   Kurzweil's assertion that the Singularity will enable us to transcend the limitations of our bodies and brains comes from Ray Kurzweil, *The Singularity Is Near* (London: Duckworth, 2005).

286   For more on quantum computers see Brian Clegg, *The God Effect: Quantum Entanglement, Science's Strangest Phenomenon* (New York: St. Martin's Press, 2006).

286   Kurzweil's prediction that the Singularity will arrive in the 2040s is from *The Singularity Is Near.*

288   Descriptions of the capabilities of modern robots and the DARPA Challenge are from Lee Gutkind, *Almost Human: Making Robots Think* (New York: W. W. Norton, 2006).

289   Damien Broderick's description of the increasing speed of human travel is covered in Bill McKibben, *Enough: Staying Human in an Engineered Age* (New York: Henry Holt, 2003).

Index